Robot Learning
from Human Teachers

Synthesis Lectures on Artificial Intelligence and Machine Learning

Editor
Ronald J. Brachman, *Yahoo! Labs*
William W. Cohen, *Carnegie Mellon University*
Peter Stone, *University of Texas at Austin*

Robot Learning from Human Teachers
Sonia Chernova and Andrea L. Thomaz

ISBN: 978-3-031-00442-1   paperback
ISBN: 978-3-031-01570-0   ebook

DOI 10.1007/978-3-031-01570-0

A Publication in the Springer series
SYNTHESIS LECTURES ON ARTIFICIAL INTELLIGENCE AND MACHINE LEARNING

Lecture #28
Series Editors: Ronald J. Brachman, *Yahoo! Labs*
           William W. Cohen, *Carnegie Mellon University*
           Peter Stone, *University of Texas at Austin*
Series ISSN
Print 1939-4608   Electronic 1939-4616

Robot Learning from Human Teachers

Sonia Chernova
Worchester Polytechnic Institute

Andrea L. Thomaz
Georgia Institute of Technology

SYNTHESIS LECTURES ON ARTIFICIAL INTELLIGENCE AND MACHINE LEARNING #28

ABSTRACT

Learning from Demonstration (LfD) explores techniques for learning a task policy from examples provided by a human teacher. The field of LfD has grown into an extensive body of literature over the past 30 years, with a wide variety of approaches for encoding human demonstrations and modeling skills and tasks. Additionally, we have recently seen a focus on gathering data from non-expert human teachers (i.e., domain experts but not robotics experts). In this book, we provide an introduction to the field with a focus on the unique technical challenges associated with designing robots that learn from naive human teachers. We begin, in the introduction, with a unification of the various terminology seen in the literature as well as an outline of the design choices one has in designing an LfD system. Chapter 2 gives a brief survey of the psychology literature that provides insights from human social learning that are relevant to designing robotic social learners. Chapter 3 walks through an LfD interaction, surveying the design choices one makes and state of the art approaches in prior work. First, is the choice of input, how the human teacher interacts with the robot to provide demonstrations. Next, is the choice of modeling technique. Currently, there is a dichotomy in the field between approaches that model low-level motor skills and those that model high-level tasks composed of primitive actions. We devote a chapter to each of these. Chapter 7 is devoted to interactive and active learning approaches that allow the robot to refine an existing task model. And finally, Chapter 8 provides best practices for evaluation of LfD systems, with a focus on how to approach experiments with human subjects in this domain.

KEYWORDS

Learning from Demonstration, imitation learning, Human-robot Interaction

Contents

CHAPTER 1

Introduction

Machine Learning techniques have had great success in many robotics applications. Today's robots are able to study the depths of the Earth's oceans, carry equipment while following soldiers through mountainous terrain, and explore the peaks and valleys of Mars. Robots build (and will soon drive) our cars, gather the items for shopping orders in busy warehouses, and keep hospital shelves stocked with supplies. Robots can vacuum your floor, mow your lawn, and clean your pool. Yet robots, or more specifically the algorithms that control them, are still unable to handle many of the complexities of the real world. Today, and for the foreseeable future, it is not possible to go to a store and bring home a robot that will clean your house, cook your breakfast, and do your laundry. These everyday tasks, while seemingly simple, contain many variations and complexities that pose insurmountable challenges for today's machine learning algorithms.

What separates impossible domains from challenging-yet-achievable ones for today's autonomous technologies is the degree of structure and consistency within the problem domain. Vacuuming robots require a flat floor to operate and not much else. Since the vast majority of house floors meet this requirement, the deployment of robotic vacuum cleaners has been highly successful. The new owner of such a robot simply needs to press the *Clean* button and in most cases the robot performs its function as expected (until it gets stuck on that sock you left on the floor, but that's why it's not a room cleaning robot).

Now consider the scenario of bringing a new house cleaning robot home, putting it in the kitchen, and pressing a similar *Clean* button for the first time. Some of the tasks such a robot might be expected to do is to load the dishwasher with all the dirty dishes, toss waste into the trash, and return clean items to their correct locations. The level of complexity of these tasks is higher not only in terms of perception and manipulation capabilities, but also in the required degree of adaptation to the new environment. Each house is unique, with custom layouts, preferred object locations, and rules (e.g., "never put the knife with the red handle in the dishwasher or it will rust"). Just as a human house guest arriving to a home for the first time, the robot needs to adapt to the customs of a particular household. This means that a single *Clean* button is no longer sufficient for such a system, instead the platform, and its underlying algorithms, must support the ability for the user to customize the robot's policy of behavior.

Robot *Learning from Demonstration (LfD)* explores techniques for learning a task policy from examples provided by a human teacher. The field of LfD has grown into an extensive body of literature over the past 30+ years, with a wide variety of approaches for encoding human demonstrations and modeling skills and tasks. In this book we provide an introduction to the field with a focus on the unique technical challenges associated with designing robots that learn from human instruction. The book is written for AI researchers who are interested in developing Learning from Demonstration methods, or who would like to learn more about different modes of interaction or evaluation methods for such systems.

1.1   MACHINE LEARNING FOR END-USERS

The above household robot scenario describes one possible application area for LfD techniques. More generally, this scenario is motivated by the challenge of enabling a novice user, a non-programmer, to customize existing robot behaviors or develop new ones through intuitive teaching methods. The motivation for tackling this challenge centers on the belief that it is impossible to pre-program all the necessary knowledge into a robot operating in a diverse, dynamic and unstructured environment. Instead, end-users must have the ability to customize the functionality of these robotic systems. Since it is impractical to assume that every end-user will have programming experience, natural and intuitive methods of interactions must be developed to enable non-roboticists to effectively use such systems.

LfD techniques build upon many standard Machine Learning methods that have had great success in a wide range of applications. However, learning from a human teacher poses additional challenges, such as limited human patience and inconsistent user input. Traditional Machine Learning techniques have not been designed for learning from ordinary human teachers in a real-time interaction, resulting in a need for new, or modified, methods. Figuring out at which level of the algorithm to involve the user is also a challenge, with different approaches being applicable to different aspects of the learning problem. Some of the design choices that go into structuring a learning problem include the following.

- *Data collection.* In any Supervised Learning process, collecting the training and testing data sets is critical to a successful learning process. The data must be representative of the states and actions that the robot will encounter in the future. The size and diversity of the training and testing data set will determine the speed and accuracy of learning and the quality of the resulting system, including its generalization characteristics. How can the teacher decide what training data to include? Can the robot make the selection or influence the decision process?

- *Selecting the feature space and its structure.* Deciding what input features and similarity metrics are most important for discriminating in the task and environment at hand is a critical step. The designer must be careful to include input features that are in fact discriminatory and the algorithm will learn faster if the redundant or non-discriminatory features are excluded. Who is responsible for performing feature selection for learning a new task through LfD?

- *Defining a reward signal.* In many learning systems, such as Reinforcement Learning (RL) [245], the reward function serves a central role in the learning process. How can the teacher effectively define a reward or objective function that accurately represents the task to be learned?

- *Subtasking the problem.* Learning speed can often be dramatically improved by splitting a task into several less complicated subtasks, although determining the subtask structure can be challenging in some domains. Should the teacher determine the task structure, or will it be determined automatically by the robot? Can the robot guide the teacher's choices and provide feedback?

These are some of the design choices that developers face in implementing interactive machine learning methods. While in many cases the answers to these questions are predetermined by the target application domain, in other situations the choice is left up to the developer.

Additionally, it's important to note that working with novice users is not the only motivation for LfD, some techniques are designed specifically with expert users in mind. Most such application areas focus on techniques for generating control strategies that would be very difficult or time consuming to program through traditional means, such as when the dynamics of the underlying system are poorly modeled or understood. In this scenario the user is often assumed to be at the very least a trained task expert, if not a roboticist. Potential application areas include a wide variety of professional fields, including manufacturing and the military.

1.2   THE LEARNING FROM DEMONSTRATION PIPELINE

Regardless of whether the target user is a novice or an expert, all Learning from Demonstration techniques share certain key properties. Figure 1.1 illustrates the LfD pipeline. This is an abstract oversimplification, but is a useful abstraction with which to frame the design process for building an LfD system. In this book, we explore the field of Learning from Demonstration from both algorithmic and Human-Robot Interaction (HRI) perspectives, by stepping through each stage of this pipeline.

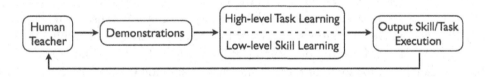

Figure 1.1: A simplified illustration of the Learning from Demonstration pipeline. This also serves as a roadmap for this book, in which chapters are devoted to each stage of the pipeline.

The assumption in all LfD work is that there exists a *Human Teacher* who demonstrates execution of a desired behavior. In Chapter 2, we consider the learning process from the human's point of view. We look at the social learning mechanisms used by humans, particularly children, in order to gain possible insights into how LfD systems might be developed and to better understand how learning robots might one day fit within the established human social norms. Then in Chapter 3, we address the *Demonstrations* component, reviewing common modes of human-robot interaction that are used to provide demonstrations.

The learner is provided with these demonstrations, and from them derives a policy—a mapping from perceived state to desired behavior—that is able to reproduce the demonstrated behavior. The ability to *generalize* across states is considered critical, since it is impractical, and often impossible, for the teacher to demonstrate the correct behavior for every possible situation that the robot might encounter. Our goal in this book is to present an overview of state of the art techniques for this policy derivation process. We do this by organizing the field into those algorithms focused on *Low-level Skill Learning* (Chapter 4) and those focused on *High-level Task Learning* (Chapter 5).

In Chapter 6 we address the ways in which this process can be made into a loop, such that an initially learned model is further refined. The ability to perform incremental learning or refinement over time, as well as the ability to generalize from a small number of demonstrations will be crucial in many domains. Factors such as the interpretability or transparency of the policy, and techniques for enabling the user to understand what knowledge the robot possesses and why it behaves in the way it does will be critical to the success of LfD methods in real-world applications.

After stepping through each aspect of the LfD pipeline, in Chapter 7 we turn the focus to evaluation. In particular, we argue for the importance of validating LfD algorithms with HRI studies. As such, this chapter contains guidelines for conducting such experiments to evaluate LfD methods with end-users. Finally, Chapter 8 is a discussion of where we see the field heading, and what we consider the most crucial future work in this exciting field.

1.3   A NOTE ON TERMINOLOGY

This book builds on an extensive collection of research literature, and one of the goals of the book is to familiarize the reader with many of the seminal works in this area. Within this research literature, LfD techniques are described by a variety of terms, such as Learning by Demonstration (LbD), Learning from Demonstration (LfD), Programming by Demonstration (PbD), Learning by Experienced Demonstrations, Assembly Plan from Observation, Learning by Showing, Learning by Watching, Learning from Observation, behavioral cloning, imitation and mimicry. While the definitions for some of these terms, such as imitation, have been loosely borrowed from other sciences, the overall use of these terms is often inconsistent or contradictory across articles. Within this book, we refer to the general category of algorithms in which a policy is derived based on demonstrated data as Learning from Demonstration, and we reference other terms as appropriate in the coming chapters.

CHAPTER 2

Human Social Learning

When a machine learner is in the presence of a human that is motivated to help, social interaction can be a key element in the success of the learning process. Although robots can also learn from observing demonstrations not directed at them, albeit less efficiently, the scenario we address here is primarily the one where a person is explicitly trying to teach the robot something in particular.

In this chapter, we review some key insights from human psychology that can influence the design of learning robots. We focus our discussion on findings in *situated learning*, a field of study that looks at the social world of a child and how it contributes to their development. In a situated learning interaction, a good instructor maintains a mental model of the learner's understanding and structures the learning task appropriately with timely feedback and guidance. The learner contributes to the process by expressing their internal state via communicative acts (e.g., expressing understanding, confusion, attention, etc.). This reciprocal and tightly coupled interaction enables the learner to leverage from instruction to build the appropriate representations and associations.

The situated learning process stands in contrast to typical scenarios of machine learning which are often neither interactive nor intuitive for a non-expert human partner. Since social learning mechanisms used by humans are both proven to be effective and naturally occurring across society, enabling robots to engage in social interaction with the user can lead to more flexible, efficient, personable and teachable machines that more closely match the user's expectations in behavior.

It is worth noting that despite its reliance on human teachers, the field of Learning from Demonstration has not focused much attention on the interactivity of the learning system. As we will see in Chapters 4 and 5, it is quite typical to first collect demonstrations in batch and then

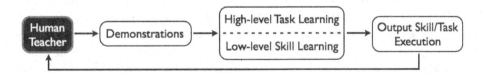

Figure 2.1: In this chapter we start with a look at the Human Teacher component of the LfD pipeline. A survey of human social learning provides insight into biases and expectations that a human may bring to the LfD process.

Figure 2.2: Starting at an early ages, children use the information around them to learn from observation, experience, and instruction, striving to imitate the adults around them.

have a learning algorithm use this data to model a skill or task later. What the work highlighted in this chapter points out is the distinction between a typical batch process and the interactivity of a social learning process. We will return to this topic in Chapter 6, where we consider how to make an LfD process interactive through online learning, high level critiques of the robot's exploration, and the incorporation of Active Learning.

In this chapter, we highlight characteristics of human social learning in the first three sections. We look at human motivation for learning, how human teachers scaffold the learning process, and what feedback human learners provide. All of these topics have implications for the technical design of robot learners, which are the focus of the remaining chapters of this book (Figure 2.1).

2.1   LEARNING IS A PART OF ALL ACTIVITY

In most Machine Learning scenarios, learning is an explicit activity. The system is designed to learn a particular thing at a particular time. With humans, on the other hand, there is an ever-present motivation for learning, a drive to improve oneself, and an ability to seek out the expertise of others. Some inspiring characteristics of a motivated learner include: a curiosity about new environments and experiences; the ability to recognize and exploit good sources of information, and to adopt such an information source as a role model; the desire to "be more like" that role model, which underlies all activity; and a sense of one's level of mastery with acquired skills, further driving the motivation to explore and learn about the world at opportune times.

Self-Determination Theory seeks to understand the mechanisms behind both intrinsic and extrinsic motivation in human behavior in general [224]. Here our focus is on situated learning interactions rather than self-motivated learning. We summarize two types of human motivation that lay the foundation for social learning interactions.

Motivated to Interact

A critical part of learning is gaining the ability to exploit the expertise of others [203]. Children put themselves in a good position to learn new things by being able to recognize, seek proximity to, and interact with their caregivers. They assume that the caregiver has their best interest in mind and even very young infants use this to their advantage when faced with an unknown situation [219].

The ability and desire to engage, communicate, and interact with others is seen from an early age. By the time infants are two months old, they can actively engage in communicative interactions or turn-taking routines with adults. Studies have shown that infants can start and stop communication with their mother through gesture and gaze, and that it is the infants that control the pace of the turn taking interaction [130, 257]. This turn taking capability is the foundation of many situated learning activities, and is a precursor to more sophisticated interactions, such as imitation. For example, Arbib characterizes learning as *assisted imitation*, a dynamic turn-taking activity [274]. Bruner characterizes social scaffolding interactions in general as asymmetric cooperation that becomes symmetric over time [99]. Thus, turn-taking engagements are an underlying framework in which learning takes place.

Turn-taking abilities are characteristically based on causal assumptions about the world. There is an expectation that the world, and particularly other actors in the world, will have some contingent response to one's activity. Thus, the ability to take advantage of these social interactions requires a robot to have models of engagement, turn taking, and other fundamental social skills. A growing body of research within the HRI field has focused on models for engagement and turn-taking. The work of [218] and [110] identifies and generates "connection events" in order for a robot to maintain engagement with a human interaction partner. Other systems have been developed to control multimodal dialog for social robots, such as the work of [128] that controls dynamic switching of behaviors in the speech and gesture modalities, and the framework of [185] that controls task-based dialog using parallelized processes with interruption handling. The work of [62] and [63] centers on building autonomous robot controllers for successfully engaging in human-like turn-taking interactions, with a computational model for regulating the speaking floor that explicitly represents and reasons about all four components of the behavior regulation problem: seizing the speaking floor, yielding the floor, holding the floor, and auditing the owner of the floor.

Motivated to Learn

Another important influence on human learning is the idea of a "like-me" bias—the propensity and ability to map between actions seen by others and done by self is seen at a very early age [174]. As the child grows older, interacting with adults, they come to understand that the adult is "like-me" and is therefore a source of information about actions and skills [274]. For example, both Bruner and Leontiev indicate that play is intrinsically motivated and that the object of play is the desire to be like adults and participate in the adult world [107]. Lave and Wenger make a similar argument for the motivation of learning altogether [155]. They develop of theory of "Legitimate Peripheral Participation," in which the driving force for learning a new practice is the learner's motivation to form their identity and become a full participant in the practice. On a large scale this is the motivation of all learning, children "wanting to become full participants in the adult world."

Litowitz has a similar explanation: the child wishes to be like the adult and is thus motivated to imitate and be lead through activities by the adult. He goes one step further, however, and poses an elegant theory of why the process stops. The child gets out of the subordinate learner role and becomes capable on its own through the very same mechanism. The desire to be like the adult extends to the meta-activity level, the child comes to want to have the adult-role of structuring activity (wanting to choose the clothes they wear, resisting being told what to do, etc.) [163].

Given this motivation to imitate, there are several ways in which an adult's behavior can influence a child's exploration or learning process. The following four social learning mechanisms have been identified in both human and animal learners [56, 254].

- *Stimulus (local) enhancement* is a mechanism through which an observer (child, novice) is drawn to objects others interact with. This facilitates learning by focusing the observer's exploration on interesting objects—ones useful to other social group members.

- *Emulation* is a process where the observer witnesses someone produce a particular result on an object, but then employs their own action repertoire to produce the result. Learning is facilitated both by attention direction to an object of interest and by observing the goal.

- *Mimicking* corresponds to the observer copying the actions of others without an appreciation of their purpose. The observer later comes to discover the effects of the action in various situations. Mimicking suggests, to the observer, actions that can produce useful results.

- *Imitation* refers to reproducing the actions of others to obtain the same results with the same goal.

Cakmak et al. [46] present an implementation of these four social learning mechanisms and articulate the distinct computational benefits of each. Their results show that all four social strategies provide learning benefits over self exploration, particularly when the target goal of learning is a rare occurrence in the environment. The work characterizes the differences between strategies, showing that the "best" one depends on both the nature of the problem space and the current behavior of the social partner.

The general concept of motivation has also been studied in the context of reinforcement learning. Intrinsically motivated RL been proposed as a framework within which agents exploit "internal reinforcement" that rewards novel situations or experiences [65, 233]. A number of other techniques for integrating self-motivation and curiosity have also been studied within the context of developmental learning [121, 200, 229], however these methodologies have not yet been applied in the context of interactive learning agents or LfD.

(a) Attention Direction   (b) Dynamic Scaffolding

Figure 2.3: Examples of scaffolding the learning process through attention direction and simplification of the task or environment.

2.2   TEACHERS SCAFFOLD THE LEARNING PROCESS

An important characteristic of a good learner is the ability to learn both on one's own and by interacting with another. Children are capable of exploring and learning on their own, but in the presence of a teacher they can take advantage of the social cues and communicative acts provided to accomplish more. For instance, the teacher often guides the child's search process by providing timely feedback, luring the child to perform desired behaviors, and controlling the environment so the appropriate cues are easy to attend to, thereby allowing the child to learn more effectively, appropriately, and flexibly. *Scaffolding* is the process by which an adult organizes a new skill into manageable steps and provides support such that a child can achieve something they would not be able to accomplish independently [99, 265]. A good teacher will scale instruction appropriately and create a good environment for learning the task at hand. In robotics, the human may be able to help the robot with hard problems like "what to learn," "when to learn," "what action to try," and "how to measure success" [35].

2.2.1   ATTENTION DIRECTION

Attention direction is one of the essential mechanisms that contributes to the learning process [268, 274]. Analyzing parent-child tutoring sessions reveals a number of ways that adults provide structure and guide attention to let children succeed: placing important objects close to the child's face, arranging the physical environment such that the desired action is within reach, or doing a demonstration in the infant's line of sight to introduce object affordances.

The adult is also implicitly directing the child's attention with their gaze direction. The tendency to follow eye gaze is seen very early on, this is a first step to reference and joint attention. It has also been shown that in order to hold joint attention and direct the infant's attention, a communicative situation must first be established. This can be with a period of eye contact, verbal, or behavioral contingent responses [76].

Within HRI research, a growing body of work has focused on social gaze behavior [117, 127, 153, 181, 182, 230, 256, 270], for example in the use of gaze for regulating turn-taking in two-party [153, 270] and multi-party conversations [24, 171, 182, 256]. These studies provide strong evidence that gaze cues from a robot support conversational functions and result in a more natural interaction with a human. As an example of applying this to context of learning, [183] showed how using human-like visual saliency detection may help a robot learner segment a teaching demonstration into steps, and determine the right aspects of the state to pay attention to during the demonstration.

Another way of directing attention is to emphasize or exaggerate parts of the desired movement. This form of instruction is challenging to adapt to LfD because the goal is not to reproduce the exaggeration itself, but instead to direct the focus of attention during learning.

2.2.2   DYNAMIC SCAFFOLDING

Dynamic scaffolding is the notion that adults create a learning situation that is the right level of complexity for the learner. The adult adjusts dynamically to make sure the child is working within the Zone of Proximal Development, defined as the gap between what a learner has already mastered and what he or she can achieve with the aid of a teacher. In a way, the teacher creates "microworlds" for the learner to master parts of the task in isolation before moving on, providing safety and intermediate attainable goals [42]. For example, with language parents first treat anything as conversational speech, but eventually they raise their expectations, scaffolding the child's conversational abilities [257]. In book reading, the parent will at first ask and answer their own questions, and later they will expect the child to participate in the question/answer game.

Closely related to this idea is Lave and Wenger's theory of legitimate peripheral participation, which states that the best way to learn is by starting on the sidelines and gradually gaining responsibility. This limits the opportunity for failure while still letting the newcomer play a legitimate part in the community. The level of scaffolding provided is an important factor in learning, instructors that always intervene to prevent problems may actually inhibit learning and the development of abilities to detect and prevent errors [219].

The idea of scaffolding has been adapted into machine learning, and LfD specifically. Several LfD techniques have leveraged the human teacher in spacial scaffolding, in which the teacher restructures the learning environment to direct or focus the attention of the learner on the most relevant aspects of the task being learned [26, 227, 228]. Within other techniques, scaffolding is used as a means to build complex behaviors by combining or adapting simpler previously taught skills [13, 14, 129].

2.2.3   EXTERNALIZING AND MODELING METACOGNITION

When working with children, adults often externalize the thinking process [23, 57]. In problem solving, a common simplification is to switch from an open-ended "wh" question (where, who, why, etc.), to yes/no questions when the child is having trouble. For example when asking "do you know where X is?" and the child says "no" or has trouble, the adult will switch to yes/no questions like "is it ... ?" to frame the search space. Often the yes/no questions are absurd to define the extremes of the space, instead exemplifying the process that the child should be using to come up with the answer for the question.

Greenfield also observes that if a child turns to an adult during a task, the adult may ask a question or give a gesture hint. The questions asked are meant to elicit the thinking process. Additionally, an important role that the adult plays in a child's learning process is linking new information to old, showing or suggesting to the child similarities between new problems and old ones [219]. A good teacher makes the information in a new problem compatible with what is known, guiding the generalization process, helping the child apply skills across various contexts.

Importantly, in humans, the key element that enables the above techniques to be successful is meta-learning. Children can go from being directed in a task through leading questions and hints to internalizing that process and being able to achieve the task on their own. Thus, in robots, it is important to not only follow instructions and model the specific activity, but to learn task strategies (e.g., questions to ask, what to pay attention to, etc.), from these interactions.

2.3   ROLE OF COMMUNICATION IN SOCIAL LEARNING

2.3.1   EXPRESSION PROVIDES FEEDBACK TO GUIDE A TEACHER

To be a good instructor, one must maintain a mental model of the learner's state (e.g., what is understood so far, what remains confusing or unknown) in order to appropriately structure the learning task with timely feedback and guidance. The learner helps the instructor by expressing their internal state via communicative acts (e.g., expressions, gestures, or vocalizations that reveal understanding, confusion, attention, etc.). Through reciprocal and tightly coupled interaction, the learner and instructor cooperate to aid both the instructor's ability to maintain a good mental model of the learner, and the learner's ability to leverage from instruction to build the appropriate models, representations, and associations.

With this view of learning as a tightly coupled collaboration, theories of human cooperative and collaborative activity help inform the design of robot learners. Cohen et al. analyzed task dialogs in which an expert instructed a novice assembling a physical device, and found that much of task dialog can be viewed in terms of joint intentions [72]. Their study identified key discourse functions including: organizational markers that synchronize the start of new joint actions ("now," "next," etc.), elaborations and clarifications for when the expert believes the apprentice does not understand, and confirmations establishing the mutual belief that a step was accomplished. Another important work is that of Bratman, in which he defines prerequisites for an activity to be considered shared and cooperative, stressing the importance of mutual responsiveness, commitment to the joint activity and commitment to mutual support [34]. Cohen et al. support these guidelines and also predict that an efficient and robust collaboration scheme in a changing environment needs an open channel of *communication*.

These theories argue for the importance of sharing information through communication in order to maintain a successful collaborative activity. Thus, a robot learner that people will find collaborative and cooperative, must take into account nonverbal communication, such as gestures and gaze, to facilitate the interaction and maintain an understandable transparent interface between the human and the machine.

2.3.2   ASKING QUESTIONS

In developmental psychology, the role of curiosity and inquiry is highlighted time and again as a crucial component to the learning process. Early in development this is characterized in self-learning where there is an active process of effectively asking questions of the environment. Piagetian self-regulatory reflexes (e.g., sucking, grasping, circular reactions) are crucial to early learning, helping infants/children obtain developmentally appropriate experiences for learning [207]. The work of Gopnik has additionally shown that children (and adults) are highly efficient in this process. In one study, Gopnik and colleagues demonstrated to children a "blicket machine" that made a sound when certain objects were put near it but not others. When asked to figure out how to make it go, they observed that 2, 3, and 4-year olds would efficiently explore the environment with actions (interventions) to uncover the pattern of conditional dependence between objects and the sound, inferring the causal structure of the machine [97].

Later, children become experts in actively seeking knowledge from their social environment, first becoming proficient at deciding to whom to pay attention. Movellan showed that children are highly efficient in their behavior, and in the face of deciding whether or not someone or something is reacting contingently to themselves, optimize their actions to gain the most information [178]. Thus, even pre-verbal children that cannot "ask questions" in the traditional sense of the term, are not passive observers but active learners in their world.

Educational psychology gives another view, looking at questions in a pedagogical context. Grasser and Person studied tutoring sessions in both grade school and college students, classifying a variety of question categories, under two main groups, those requiring short answers vs. long answers. They then studied the frequency and intent of various questions in real tutorial settings. They found the frequency of different types of questions was similar across two different settings, and that students primarily ask questions because of a knowledge deficit and to maintain common ground (e.g., confirming knowledge) [98]. In other research they have shown that the quality of a student's questions and the completeness of their answers are the best predictors of final exam performance. Hence, performance was not correlated with answers students gave to confirming questions like "did you get that" [204]. Thus, a good teacher must do more than ask for knowledge confirmations to maintain a good mental model of the learner's current knowledge.

Figure 2.4: Simon, at Georgia Tech, is one example of a robot designed with both learning and social interaction in mind. Techniques for making use of scaffolding, attention direction, transparency, and question asking are central to the development of this system.

These experiments quantifying question usage are closely related to HRI goals, and techniques integrating some of these principles into LfD will be discussed in Chapter 6.

2.4 IMPLICATIONS FOR THE DESIGN OF ROBOT LEARNERS

The human learning process serves as an inspiration in the design of social learning robots. By studying human learning we gain insights into the design of advanced learning systems. Furthermore, because learning from demonstration inherently requires interaction between the robot and the user, designing the interaction to conform to the user's expectations leads to a more natural and effective learning process. The extent to which social elements need to be integrated into LfD often depends on the application. In some circumstances, the robot may benefit from the full range of social interactions, taking into account social cues such as gestures, gaze, direction of attention, and possibly even extending to affect. In other applications, minimal or no social understanding may be required, with the interaction instead limited to a human-computer interface. In all cases, the designers of the robot strive for the most natural, flexible, and efficient learning system for the given task. The following design elements are some that should be considered in the design of robots that learn from demonstration.

- **Social interaction.** Should the robot leverage the social aspect of the interaction? Would learning be aided if the robot understood the social cues of the user? Would learning be aided if the robot could exhibit social cues? Which social cues are most effective for LfD interactions? Which social cues, whether from the robot or teacher, are most informative for task learning, and which social cues are most preferred by users?

- **Motivation for learning.** Does the robot require intrinsic motivation for learning, or will all learning be initiated and directed by the human user?

- **Transparency.** To be effective, a teacher must be able to maintain an as accurate a mental model of the learner's knowledge as possible. How can the robot externalize what it has learned and make elements of the internal model transparent to the user? What techniques for communicating the learner's knowledge should be used to aid the learning process? Is it necessary that the communication techniques mimic the way humans communicate, or is it equally (or more) effective to leverage interfaces that are not part of natural human communication, such as screen-based devices?

- **Question asking.** Asking questions is a critical part of the human learning process. How does the robot effectively communicate the limits of its knowledge or pose a question? How can the user frame the answer in a way that the robot can understand, and how should the gained information be used to improve the underlying model? Many different types of questions can be designed, such as "what should I do now?" or "what is the intended goal?" Given multiple possible questions, how can the robot determine which questions to ask?

- **Scaffolding.** Just as for humans, complex tasks can be easier for machines to learn if they are broken down into simpler components. Organization of knowledge or skills into simpler parts also often allows for greater efficiency through reuse. How can the robot leverage scaffolding in its learning and interaction with the user? How can previously learned policies be built upon and reused in new settings? Note that in addition to simply saving learned policies, this could involve *parameterizing* the action space of the robot, allowing a previously learned skill (e.g., pick up box) to generalize to new objects or scenarios.

- **Directing attention.** Humans use a number of techniques to control the direction and scope of attention within a conversation. In the context of learning, both in the role of a teacher and a student asking a question, this skill is often used to focus learning, akin to feature selection in machine learning. How can control of attention be leveraged to simplify learning in complex domains? How can the robot direct the attention of the user, and vice versa? How does the learning algorithm respond to shifts in attention?

- **Online vs. batch learning.** The majority of traditional machine learning techniques make use of a batch learning process, examining all the training data at once and producing a model. Learning from demonstration can be cast as a batch learning process that occurs at the end of a training session, or once enough new demonstrations are acquired. However, it can also be viewed as an online learning process in which training data is acquired incrementally, similar to active learning. The choice between online and batch learning is important in the design of an interactive learning system as it will determine the flow of interaction and how new training data is acquired and integrated into the model.

As can be seen from this discussion, social learning mechanisms have the potential to play an important role in every part of the LfD process. In the next chapter, and the ones that follow, we switch to looking at LfD from a computational perspective, studying the Machine Learning techniques that can be applied to this problem. However, human involvement remains a critical factor in the discussed methods, and we return to this topic in Chapter 6, where we consider interactive techniques for policy refinement.

CHAPTER 3

Modes of Interaction with a Teacher

With insights from human social learning in mind, in this chapter we turn to a central design choice for every Learning from Demonstration (LfD) system: how to solicit demonstrations from the human teacher. As highlighted in Figure 3.1, this chapter forms the introduction to the technical portion of the book, laying the foundation for the discussion of both high-level and low-level learning methods. We do not entirely ignore the issues of usability and social interaction, after all, the choice of interaction method will impact not only the type of data available for policy learning, but also many of the topics discussed in the previous chapter (e.g., transparency, question asking, directing attention). However, these topics will remain in the background until Chapters 6 and 7, in which we discuss policy refinement and user study evaluation, respectively.

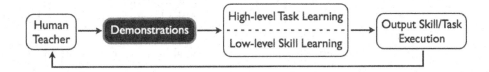

Figure 3.1: In this chapter, we discuss a wide range of techniques for collecting demonstration input for LfD algorithms.

In this chapter, we first introduce readers to the correspondence problem, which pertains to the differences in the capabilities and physical embodiment between the robot and user. We then characterize demonstration techniques under three general modes of interaction, which enable a robot to learn through *doing*, through *observation*, and from *critique*.

3.1 THE CORRESPONDENCE PROBLEM

An LfD dataset is typically composed of state-action pairs recorded during teacher executions of the desired behavior, sometimes supplemented with additional information. Exactly how demonstrations are recorded, and what the teacher uses as a platform for the execution, varies greatly across approaches. Examples range from sensors on the robot learner recording its own actions as it is passively teleoperated by the teacher, to a camera recording a human teacher as she executes

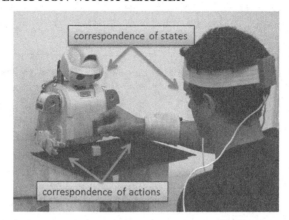

Figure 3.2: The correspondence problem arises due to the differences in the sensing abilities and physical embodiment between the human and robot, making it more challenging to accurately map between their respective state and action representations [49].

the behavior with her own body. Some techniques have also examined the use of robotic teachers, hand-written control policies and simulated planners for demonstration.

For LfD to be successful, the states and actions in the learning dataset must be usable by the learner. In the most straightforward setup, the states and actions recorded during the demonstrations map directly to the sensing and movement capabilities of the robot. In other cases, however, a direct mapping does not exist between the teacher and learner due to differences in sensing ability, body structure or mechanics. For example, a robot learner's camera will not detect state changes in the same manner as a human teacher's eyes, nor will its gripper apply force in the same manner as a human hand. The challenges which arise from these differences are referred to broadly as the *correspondence problem* [186]. Specifically, the issue of correspondence deals with the identification of a mapping between the teacher and the learner that allows the transfer of information from one to the other.

The correspondence problem lies at the heart of Learning from Demonstration, and is intertwined in the choice of both the human-robot interaction method and computational technique used for learning. Using a direct demonstration technique that does not require correspondence simplifies the learning process significantly as it removes one source of possible error—the mapping function that translates human capabilities to those of the robot. As discussed below, several demonstration techniques directly map between the actions of the teacher and those of the student, the primary examples of which are teleoperation of the robot through kinesthetic teaching [51] or a controller such as a joystick or computer interface [1, 237]. However, not all systems are amenable to teleoperation. For example, low-level motion demonstrations are difficult on systems with complex motor control, such as high degree of freedom humanoids. Furthermore,

(a)   (b)

Figure 3.3: (a) Kinesthetic teaching with the iCub robot [13]. (b) User controlling the full-body motions of an Aldebaran Nao robot using the Xsens MVN inertial motion capture suit [141].

physically controlling the robot may not be natural, or even possible, in a given situation. Instead, the teacher may find it more effective to perform the task with their own body while the robot watches. Enabling the robot to learn from observations of the teacher requires a solution for the correspondence problem, the states/actions of the teacher during the execution must be to be inferred and mapped onto the abilities of the robot. Learning in such settings depends heavily upon the accuracy of this mapping. Finally, the teacher may not demonstrate the task at all, and instead observe the robot and provide critique or corrections to the current behavior. In the following sections we discuss techniques for enabling the robot to learn from its own experiences, observation of the teacher and the teacher's critiques. We conclude the chapter with a discussion of the tradeoffs and implications that the choice of interaction mode has on the design of the overall robot learning system.

3.2   LEARNING BY DOING

Teleoperation provides the most direct method for information transfer within demonstration learning. During teleoperation, the robot is operated by the teacher while recording from its own sensors. Demonstrations recorded through human teleoperation via a joystick have been used in a variety of applications, including flying a robotic helicopter [1], soccer kicking motions [40], robotic arm assembly tasks [64], and obstacle avoidance and navigation [118, 237]. Teleoperation has also been applied to a wide variety of simulated domains, such as mazes [70, 214], driving [3, 66], and soccer [7], and many other applications. Teleoperation interfaces vary in complexity from hand-held controllers to teleoperation suits [159]. Hand-written controllers have also been used to teleoperate the robot in the place of a human teacher [11, 102, 221, 237].

   Kinesthetic teaching offers another variant for teleoperation. In this method, the robot is not actively controlled, but rather its passive joints are moved through the desired motions while the robot records the trajectory [51]. Figure 3.3(a) shows a person teaching a humanoid robot to manipulate an object. This technique has been extensively used in motion trajectory learning, and many complementary computational methods are discussed in Chapter 4. A key benefit of teaching through this method of interaction is that it ensures that the demonstrations are constrained to actions that are within the robot's abilities, and the correspondence problem is largely eliminated. Additionally, the user is able to directly experience the limitation of the robot's movements, and thus gain greater understanding about the robot's abilities.

Another alternative to direct teleoperation is *shadowing*, in which the robot mimics the teacher's demonstrated motions while recording from its own sensors. In comparison to teleoperation, shadowing requires an extra algorithmic component which enables the robot to track and actively shadow (rather than be passively moved by) the teacher. Body sensors are often used to track the teacher's movement with a high degree of accuracy. Figure 3.3(b) shows an example setup used by [141], in which the Xsens MVN inertial motion capture suit worn by the user is used to control the robot's pose. This example demonstrates tightly coupled interaction between the user and the robot, since almost every teacher movement is detected by the sensors.

Shadowing also allows for loosely coupled interactions, and has even been applied to robotic teachers. Hayes and Demiris [109] perform shadowing with a robot teacher whose platform is identical to the robot learner; the learner follows behind the teacher as it navigates through a maze. Nehmzow et al. [187] present an algorithm for robot motion control in which the robot first records the human teacher's execution of the desired navigation trajectory, and then shadows this execution. While repeating the teacher's trajectory, the robot records data about its environment using its onboard sensors. The action and sensor data are then combined into a feedback controller that is used to reproduce future instances of the demonstrated task.

Trajectory information collected through teleoperation, kinesthetic teaching or shadowing can be combined with other input modalities, such as speech. Nicolescu and Mataric [190] present an approach in which a robot learns by shadowing a robotic or human teacher. In addition to trajectory information, their technique enables the teacher to use simple voice cues to frame the learning ("here," "take," "drop," "stop"), to provide informational cues about the relevance or irrelevance of observation inputs and indications of the desired behavioral output. In Rybski et al. [225], demonstration of the desired task is also performed through shadowing combined with dialog in which the robot is told specifically what actions to execute in various states. Meriçli et al. [175] present a similarly motivated approach which additionally supports repetitions (cycles) in the task representation and enables the user to modify and correct an existing task. Breazeal et al. [36] also explore this form of demonstration, enabling a robot to learn a symbolic high-level task within a social dialog.

Finally, some learning methods pay attention only to the state sequences, without recording any actions. This makes it possible to communicate the task objective function to the learner without traditional action demonstrations. For example, by drawing a path through a 2-D representation of the physical world, Ratliff et al. provide high-level path planning demonstrations to

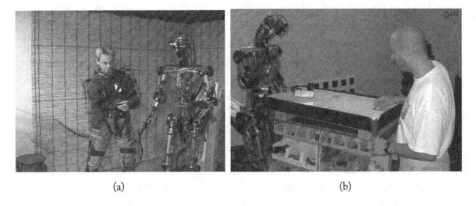

(a)   (b)

Figure 3.4: (a) User teaching a forehand swing motion to a humanoid robot using the Sarcos Sen-Suit [115]. (b) Humanoid robot learning to play air hockey from observation of opponent player [25].

a rugged outdoor robot [215] and a small quadruped robot [143, 216]. Human-controlled tele-operation demonstrations are also utilized with the same outdoor robot for lower-level obstacle avoidance [216]. Since actions are not provided in the demonstration data, at run time a learned state-action mapping does *not* exist to provide guidance for action selection. Instead, actions are selected by employing low level motion planners and controllers [215, 216], and provided transition models [143].

3.3   LEARNING FROM OBSERVATION

In many situations, it is more effective or natural for the teacher to perform the task demonstration using their own body instead of controlling the robot directly. As discussed above, this form of demonstration introduces a correspondence problem with respect to the mapping between the teacher's and robot's state and actions. As a result, this technique is commonly used with humanoid or anthropomorphic robots, since the robot's resemblance to a human results in a simpler and more intuitive mapping, though learning with other robot embodiments is also possible. Unlike in the use of shadowing, the robot does not simultaneously mimic the teacher's actions during the observation.

Accurately sensing the teacher's actions is critical for the success of this approach. Traditionally, many techniques have relied on instrumenting the teacher's body with sensors, including the use of motion capture systems and inertial sensors. Ijspeert et al. [114, 115] use a Sarcos Sen-Suit worn by the user to simultaneously record 35 DOF motion. The recorded joint angles were used to teach a 30-DoF humanoid to drum, reach, draw patterns, and perform tennis swings (Figure 3.3(a)). This work is extended in [184] to walking patterns. The same device, supplemented with Hall sensors, is used by Billard et al. to teach a humanoid robot to manipulate boxes in sequence [29]. In later work, Calinon and Billard combine demonstrations executed by human teacher via wearable motion sensors with kinesthetic teaching [50].

Wearable sensors, and other forms of specialized recording devices, provide a high degree of accuracy in the observations. However, their use restricts the adoption of such learning methods beyond research laboratories and niche applications. A number of approaches have been designed to use only camera data. One of the earliest works in this area was the 1994 paper by Kuniyoshi et al. [152], in which a robot extracts the action sequence and infers and executes a task plan based on observations of a human hand demonstrating a blocks assembly task. Another example of this demonstration approach includes the work of Bentivegna et al. [25], in which a 37-DoF humanoid learns to play air hockey by tracking the position of the human opponent's paddle (Figure 3.3(b)). Visual markers are also often used to improve the quality of visual information, such as in [30], where reaching patterns are taught to a simulated humanoid. Markers are similarly used to optically track human motion in [122, 123, 259] and to teach manipulation [209] and motion sequences [10]. In recent years, the availability of low-cost depth sensors (e.g., Microsoft Kinect) and their associated body pose tracking methods makes this a great source of input data for LfD methods that rely on external observations of the teacher (e.g., [79]).

Related to the learning by observation problem, several works focus exclusively on the perceptual-motor mapping problem of LfD, where in order to imitate the robot has to map a sensed experience to a corresponding motor output. Often this is treated as a supervised learning problem, where the robot is given several sensory observations of a particular motor action. Demiris and Hayes use forward models as the mechanism to solve the dual-task of recognition and generation of action[80]. Mataric and Jenkins suggest behavior primitives as a useful action representation mechanism for imitation [122]. In their work on facial imitation, Breazeal et al. use an imitation game to facilitate learning the sensory-motor mapping of facial features tracked with a camera to robot facial motors. In a turn-taking interaction the human first imitates the robot as it performs a series of its primitive actions, teaching it the mapping, then the robot is able to imitate [37].

Finally, observations can also focus on the effects of the teacher's actions instead of the action movements themselves. Tracking the trajectories of the objects being manipulated by the teacher, as in [249], can enable the robot to infer the desired task model and to generate a plan that imitates the observed behavior.

3.4 LEARNING FROM CRITIQUE

The approaches described in the above sections capture demonstrations in the form of state-action pairs, relying on the human's ability to directly perform the task through one of the many possible interaction methods. While this is one of the most common demonstration techniques, other forms of input also exist in addition to, or in place of, such methods.

In *learning from critique* or *shaping*, the robot practices the task, often selecting actions through exploration, while the teacher provides feedback to indicate the desirability of the ex-

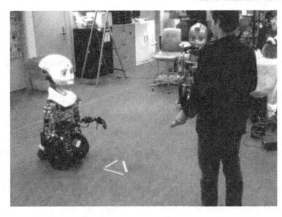

Figure 3.5: A robot learning from critique provided by the user through a hand-held remote [138].

hibited behavior. The idea of shaping is borrowed from psychology, in which behavioral shaping is defined as a training procedure that uses reinforcement to condition the desired behavior in a human or animal [234]. During training, the reward signal is initially used to reinforce any tendency towards the correct behavior, but is gradually changed to reward successively more difficult elements of the task.

Shaping methods with human-controlled rewards have been successfully demonstrated in a variety of software agent applications [33, 135, 252] as well as robots [129, 138, 242]. Most of the developed techniques extend traditional Reinforcement Learning (RL) frameworks [245]. A common approach is to let the human directly control the reward signal to the agent [91, 119, 138, 241]. For example, in Figure 3.4, the human trainer provides positive and negative reward feedback via a hand-held remote in order to train the robot to perform the desired behavior [138].

Thomaz and Breazeal extended critique-based methods to additionally allow the user to influence the selection of the next action, showing that this provides performances gains of up to 50% over a feedback-only approach [253]. Several other approaches let the human supervise an RL agent by occasionally biasing action selection rather than directly controlling all of the agent's actions [69, 148, 170]. All of these techniques have the benefit that the human need not know exactly how the agent should perform the task, and learning does not require their undivided attention. However, in many of the approaches above, one could argue that the machine does not take *enough* advantage of the human teacher that is actively willing to help. A more detailed discussion of Machine Learning techniques that leverage feedback and critique is included in Chapter 6.

3.5   DESIGN IMPLICATIONS

Given the range of input modalities seen across LfD approaches, a system designer should carefully consider which to use in a given learning domain for a given learning task. Several components should be considered in this decision.

One is the available forms of human-robot interaction. In some cases the choice of interaction modality may be restricted by environmental conditions or user factors of the target application. For example, kinesthetic teaching is ruled out if the user is not co-present with the robot, is physically unable to perform the task, or if physical contact with the robot is unsafe. In other cases, it may be possible to do kinesthetic teaching, but if the teacher is not very familiar with operating the robot or with the robot's kinematic workspace, it may be more intuitive for them to provide input by demonstrating the task themselves.

In both learning from experience and observation, an important consideration is how proficient the teacher is expected to be in demonstrating the target task. Many demonstration techniques rely only on demonstrated data to learn the task policy. In such cases, the performance of the robot can be limited by that of the teacher. It is important to remember that human demonstrations are often noisy and suboptimal in performance. As discussed in detail in Chapter 6, a number of interactive techniques exist for *refining* a policy, which help to mitigate the effects of suboptimal demonstrations. For example, exploration based techniques (e.g., RL) can be used to allow the robot to refine the task on its own, possibly in combination with teacher critique.

Additionally, demonstration data recorded by real robots frequently does not represent the full observation state of the teacher. This occurs if, while executing the task, the teacher employs extra data which is not recorded. For example, if the teacher makes decisions based on what he observes in parts of the world which are inaccessible from the robot's cameras (e.g., behind the robot, if its cameras are forward-facing). In this case, the state as observed by the teacher differs from what is actually recorded as data, sometimes making it impossible for the robot to learn the task correctly. While in many cases this is viewed simply as an additional factor to take into account, a small number of works have addressed this problem; for example, in [104] and [102] a vision-based robot is teleoperated while the teacher looks exclusively at a screen displaying the robot's camera output.

In the following chapters we study how demonstration data, collected through one of the methods described in this chapter, can be used to learn low-level or high-level policies from demonstration. As we will see, the choice of demonstration technique does little to restrict the choice of learning method, so in the future we will often talk about training sets of state-action pairs without discussing exactly the way in which they were recorded.

C H A P T E R 4

Learning Low-Level Motion Trajectories

We have thus far covered the first two stages of the LfD pipeline (Figure 4.1), looking at human social learning, as well as the range of input a LfD algorithm can be designed to work with. We now turn our attention to the wide range of algorithms for building skill and task models from demonstration data. In this chapter we focus on approaches that learn new motions or primitive actions. The motivation behind learning new motions is typically that they would be used in service of some specific task in a given domain. Hence, we will also refer to these primitive actions as *low-level motions* in contrast to the *high-level tasks* in which they would be used (covered in Chapter 5). In the literature there are several different names given to this class of "low-level" action learning, thus in this chapter we use the terms *skill, motor skill, primitive action,* and *low-level motion* interchangeably.

Figure 4.1: In this chapter we focus on approaches that learn new low-level skills, motions, or primitive actions.

The goal of learning in this context is to build an accurate model of a demonstrated primitive action, such that it could be generally applied to a variety of domain specific tasks. The notions of "accuracy" and "generality" will differ across approaches and also depend on the target domain.

Virtually all methods of supervised learning have been applied in the context of skill learning. Some seminal early examples of skill learning from demonstration include the use of Neural Networks [17], Inductive Logic [89], and Petri Nets [172] to model skills. The vast majority of modern approaches, however, are based on either Dynamic Movement Primitives (DMPs) or probabilistic modeling methods such as Hidden Markov Models (HMMs), Gaussian Mixture Regression (GMR), or some combination thereof. In this chapter, we first discuss the feature selection problem for low-level skills. We then introduce the primary approaches to skill learning, and finish with a discussion of sensitivity to sub-optimal input data.

4.1   STATE SPACES FOR MOTION LEARNING

Let us start the topic of motion learning with a discussion that is rarely addressed, the choice of state space. When surveying the field of motion learning one sees that there is a vast range of state spaces that have been employed for representing skills. In particular, the choice of state space tends to be specific to a particular target domain or skill set. Thus, it is an important design choice to consider for your LfD problem.

The easiest state for the robot learner to record and reproduce is often simply joint positions of the entire kinematic chain over time. For gestures and other free space motion this is a good state space for learning [53, 122]. However, in some cases the goal of a skill is not captured only by joint positions over time, so the state representation needs to be augmented in order to promote learning a general form of the skill. A common solution is to instead represent the same joint space trajectory as motion with respect to some target task object, also called the task frame. Take for example the skill of touching or picking up a particular object. Given multiple demonstrations of the skill with the target object in different locations, these trajectories may show very little similarity in the robot frame, but when converted to the task frame a general skill model is easily represented. For object directed skills, a further simplification on the task frame representation is often to only consider the Cartesian position of the end-effector with respect to a target object, rather than the entire kinematic chain. In this case, when reproducing a learned skill, a planner is used to determine appropriate positions for all the joints given a desired end-effector position over time.

A particular object-directed skill that has been a common target domain for LfD is assembly tasks, and several different state spaces have been used for learning primitive actions in this context. The task goal for assembly involves skills that have an end-effector make contact with objects and/or move some object in a particular way with respect to other objects (e.g., a canonical peg-in-hole skill). One way this has been addressed is by modeling the compliance of the robot's end-effectors, where the learned skill model represents the expected forces over time at either just the end-effector or in all joints of the manipulator [17, 235, 267]. An alternative is to explicitly represent the set of all possible contact states between the objects of interest [113]. The use of relational representations of contact forces has been a common approach to assembly tasks since the 1980s [88]. More modern approaches to this problem are focused on automatically extracting a skill specific discretization of the state space, e.g., the approach in [173] simultaneously estimates object contact formations and their constraints from a human demonstration of the skill. For complex dexterous manipulation tasks or learning specific grasp strategies for particular objects, a simple object contact representation may not be high enough resolution to represent the skill, forces measured at the manipulator fingertips may be necessary [162].

The goal of the representations above are to be generic, to represent any kind of motion or any kind of object directed motion. In addition to these, it is common to see state representations in LfD that are carefully constructed with perceptual information specific to a given control task, such as playing pool [202], playing tennis [180], flipping pancakes [145], or flying a helicopter [2].

In these examples the designer knows the space in which a general skill model exists and defines this for the learner. This is often seen in the case of complex dynamic skills, where even when the state space of the objective function is known, programming an optimal motor controller is nontrivial and learning from demonstration is a nice alternative.

It is also important to note that the choice of state representation has been biased by the capabilities of the hardware and platforms available. LfD has been biased toward learning skill models comprised of motion signals due to the availability of platforms for which recording and reproducing a particular motor trajectory is possible. As we see progress on robot actuators and sensors, the field of LfD will be able to address learning from other signals. For example, it is becoming increasingly common to see LfD approaches that incorporating force or compliance profiles of a skill, due to advances and availability of robots with compliant actuators.

4.2   MODELING AN ACTION WITH DYNAMIC MOVEMENT PRIMITIVES

The Dynamic Movement Primitive (DMP) framework has its roots in a paper published by Ijspeert et al. [114]. It was created to learn the attractor landscape of a controller from a single demonstration. There are various formulations of the DMP framework since its first introduction. Here we provide a basic introduction to the approach.

This approach is designed for single shot motion where the goal is to reach a particular target by the end of the skill. This is represented as a damped spring that is attached to the goal position and is modulated by a nonlinear term:

$$\ddot{x} = K(g - x) - D\dot{x} + f(\omega, s). \tag{4.1}$$

In Eq. 4.1, x, $\dot{x}$, and $\ddot{x}$ are the position, velocity and acceleration of the system (or of a single dimension of the system in a multi-DOF control policy), and g is the goal state of the skill. K is a spring constant, and D is the damping term. The first part of this equation resembles a simple PD-controller where the K and D are the *position gain* and *derivative gain* respectively. With appropriate setting of these parameters, and $f = 0$, g becomes the point attractor for the system. The parameters are either estimated or empirically chosen such that there is no overshoot if the PD portion was used alone (i.e., ideally the parameters are set such that the system is critically damped).

The f term adds acceleration to the simple point attractor system, such that it moves to the target position in a particular way. f is a nonlinear function with the parameters ω, and s is the phase variable:

$$f = \frac{\sum_i \psi_i(s)\omega_i s}{\sum_i \psi_i(s)}. \tag{4.2}$$

The variable s lets f not depend explicitly on time, rather it depends on this phase term that has either linear first-order or second-order decay dynamics (e.g., $\dot{s} = -\alpha s$) independent of the state x. Typically, this decays exponentially from 1 to 0, ensuring that the skill converges to the target point. Each dimension of the controller has an individual version of Eq. 4.1. The phase variable acts as a coupling term between them.

The usual form of f is the sum of Gaussian basis functions with s as the variable. The basis functions have distributed centers c_i throughout the [0; 1] decay zone of s; and varying widths h_i:

$$\psi_i(s) = \exp^{(h_i(s-c_i)^2)} .$$

Learning a DMP typically uses just a single skill demonstration, and involves estimating the form of f that defines how to perturb the point attractor system in the right way. Rearranging Eq. 4.1, we have the following:

$$f(\omega, s) = \ddot{x} - K(g - x) + D\dot{x}. \tag{4.3}$$

First, $x(t)$, $\dot{x}(t)$ and $\ddot{x}(t)$ are calculated from the example trajectory. The decay speed of s (e.g., the parameter α) is determined from the demonstration. Then, calculating ω given $f(s)$ is a supervised linear regression problem. One method is to assume all c basis functions have the same form (e.g., same h), evenly distributed throughout the range of s. Then the problem is reduced to learning the weight parameters ω, and a method like Locally Weighted Regression has been shown to accomplish this task [20]. There are three main parameters to tune for the DMP method: the number of basis functions in the non-linear term; the decay term α; and the position gain (derivative gain can be calculated from position gain by assuming critical damping.) For the basis functions, the distribution of centers and the variance of the widths should be done with the decay speed considered, with centers more tightly spaced in the more dynamic portions of the skill. Alternatively, the Locally Weighted Projection Regression method can be used to simultaneously learn this weight vector along with the location and width of the basis functions [264].

The set of equations (in the form of Eq. 4.1) for all the dimensions with all the parameters initialized can be treated as the model of the skill. The resulting policy can be summarized as $\dot{x} = \pi(x, t)$. Different versions of DMPs generalize the skill differently (affine transformed, scaled versions, etc.) but the assumption is typically that the skill can be taught with a single demonstration, and hence the resulting controller will closely resemble the seed demonstration.

Next, we give examples of using DMPs in practice. In [201], Pastor et al. extend the basic DMP formulation such that the initial position x_0 and the goal position g are model parameters. This enables a single DMP to handle workspace perturbations, such as moving the target object of a manipulation action to a new location. Such perturbations were demonstrated with stacking objects and pouring into a cup. Building on this, [202] is a state of the art example of LfD with DMPs, where a PR2 robot learns two different complex dynamic skills: learning a pool stroke, and learning to flip a box with chopsticks (Figure 4.2).

One place in which a system designer encodes domain knowledge into the DMP learning process is through the definition of the task space representation x. For the pool stroke skill the designers selected the five task frame variables: translational offset of the right gripper to the pool

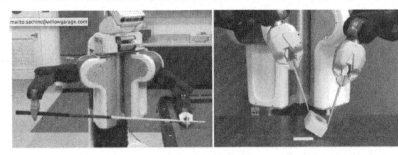

Figure 4.2: State of the art example in learning DMPs from a single demonstration, applied for the complex dynamic skills of playing pool and using chopsticks [202].

cue bridge, the roll, pitch and yaw of the cue around the bridge, as well as the elbow pose. For the chopsticks skill on the other hand the relevant variables include data from the pressure sensor array on the gripper finger tips.

Once an initial DMP model is estimated using a process similar to the basic one described above, Theodorou et al. employ a Reinforcement Learning (RL) algorithm (Policy Improvement with Path Integrals (PI^2) [251]) to further optimize the DMP estimated from demonstration. This optimization requires a cost function to be minimized, which needs to be provided by the system designer. For example, the objective of the pool stroke skill is to maximize the cue ball speed while having the ball cross the center of the scan line. With this objective function their results show that the system was able to optimize skill performance in approximately 20 min. In the chopsticks skill, the initial DMP estimated from the single successful demonstration was only able to successfully flip the box 3 of 100 runs. After the optimization step (which they report took 35 min) the robot achieved the box flip skill 172 of 200 runs. In related work, Peters and Schaal [205] compare four different such policy search methods of Reinforcement Learning in the scenario of optimizing a learned DMP motor skill. Their method, time-variant episodic Natural Actor-Critic, is shown to be the preferred method with experiments in learning a baseball swing.

DMPs are meant to be primitives used in the context of a larger task, but only a few works have pushed on this higher level aspect of the problem. One example is Mulling et al. [180] which uses a mixture of several DMPs (learned separately) to solve the larger task of playing table tennis. The specification of how to blend the control signals from two DMPs is task dependent and predefined. A related problem is to learn multiple DMPs from a single set of demonstration data. Almingol et al. [8] assume that each demonstration trajectory in the set represents one DMP, and simultaneously infer the number of primitives contained in the demonstration data as well as the DMPs themselves. Niekum et al. [193] assume that a single demonstration trajectory exhibits a sequence of motion primitives, and their approach clusters and segments trajectories in order to learn a separate DMP for each segment. Chiappa and Peters [68] also extract motion primitives from a larger time series data by detecting changes in the motion dynamics.

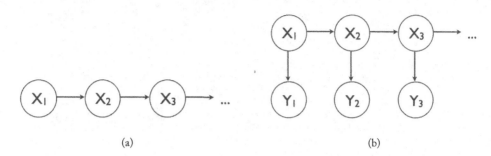

(a)   (b)

Figure 4.3: (a) First three time steps of a Markov Model for the evolution of the state variable X. The model is fully specified by the probabilities $P(X_1)$ and $P(X_t|X_{t-1})$. (b) First three time steps of a Hidden Markov Model where the observation Y is only dependent on the underlying state X, i.e., the observation probabilities $P(Y_t|X_t)$.

4.3   MODELING ACTION WITH PROBABILISTIC MODELS

The second class of approaches we introduce is those based on probabilistic models, which encompasses a large number of the prior work in skill learning. First we introduce the general approach to modeling an action with a Hidden Markov Model, the most basic instantiation of a Dynamic Bayes Net. We then discuss several different ways people have used and extended the basic approach over the years.

For a detailed introduction to the topic of HMMs the reader should refer to [211]. Here, we briefly introduce the terminology and focus our discussion on its typical usage in an LfD context. A Markov Model is a chain-structured Bayesian Network with a state variable X that evolves over time. The most commonly used form is the first-order Markov Model, which makes the simplifying independence assumption that the value of X at one point in time only depends on its value in the previous time step (Figure 4.3(a)). A Markov Model is represented by two probability distributions: the prior probability $P(X_1)$, and the transition probabilities $P(X_t|X_{t-1})$. In a Hidden Markov Model the state X is only indirectly observable through observations Y that we assume depends only on the current state (Figure 4.3(b)). Thus, in addition to the priors and transition probabilities, we now have the observation probabilities $P(Y_t|X_t)$.

In the case of skill learning from demonstration, the model of a skill is a sequence of hidden states, with prior probabilities, transition probabilities and observation probabilities. The *observation* is the continuous time series demonstration trajectory (robot joint positions, or Cartesian position of the end-effector, or position of the end-effector with respect to a target object, etc.). The problem of learning this model has two parts: structure learning and parameter learning.

Structure learning involves determining the number, k, of hidden states there are for the skill, and the connectivity of the states. A commonly used technique is to try values for k, starting with $k = 1$, and then use Bayesian Information Criterion or Minimum Description Length to select the k that best balances a tradeoff between model complexity and fit to the demonstration data. Growing HMMs is an incremental approach for learning HMM skill models [261] that explicitly deals with the structure learning problem of deciding how to set k, and changing that over time after seeing more examples of the skill. This work was done in the context of learning models of car and pedestrian behavior from video, but the approach is relevant to skill learning on robots.

Parameter learning involves estimating the three probability distributions mentioned above (also called the model λ) from the demonstrated examples of the skill. This is accomplished with the Baum-Welch algorithm [81]. Baum-Welch is an expectation-maximization (EM) algorithm that alternates between the *E-step* of computing the log likelihood of the observation sequence Y given the current model λ, and the *M-step* of adjusting the parameters in λ to maximize this likelihood. The algorithm is guaranteed to provide a monotonically increasing convergence of $P(Y|\lambda)$. To avoid EM converging to a singular solution, it is typical to set a minimum value for the eigenvalues of the covariances (or add a regularization term). This advice is even more important when training a model from a single demonstration, in order to avoid numerical problems in the estimation of the model's parameters. However, most of the works in LfD employing HMMs to represent observed trajectories exploited the use of multiple demonstrations, promoting the extraction of additional relevant information about the task, in the form of local variation and correlation information. Thus, it is more appealing to use multiple demonstrations to fully exploit the representational properties of the model.

The learned HMM can be used for recognition, as in the common application for speech or gesture recognition. Given an observation sequence, the Viterbi algorithm gives the optimal sequence through HMM states for this observation. But an HMM is also a generative model. To execute the skill, the probability distributions can be used to forward simulate a likely sequence of X and Y, i.e., the canonical way to run this action from a particular initial state Y_1. Then a controller is used to follow the determined Y trajectory. It is important to note that in practice, generating a trajectory in this way is not as simple as it sounds since Y does not necessarily form a smooth trajectory, but a discontinuous stepwise trajectory. One solution is to generate many stochastic samples and average over them, but this has the side effect of potentially smoothing important desired peaks in the movement trajectory.

The work of Kulic et al. exemplifies using HMMs for skill learning from demonstration in a robust and realistic scenario [149, 150, 151]. Figure 4.4 shows the entire pipeline of the approach. Most approaches assume that the data for a motion primitive is nicely segmented into a portion of data that represents each action that is to be modeled. Moreover it is assumed that the learning process operates over this data set of segmented examples in batch. Instead, Kulic et al. relaxes both of these assumptions and performs unsupervised segmentation of an input motion into several candidate primitive motions using the Kohlmorgen and Lemm algorithm [142] for segmentation. This results in a set of short segments that are candidate motion primitives, and an HMM is learned for each candidate. These are then clustered into a hierarchical tree structure

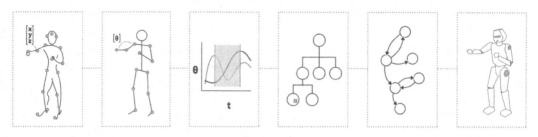

Figure 4.4: A state of the art example in learning HMMs from human demonstrations.[1] This illustrates the learning pipeline seen in the work of Kulic et al. A continuous input stream of marker position data is converted into joint angle data for a humanoid kinematic model. This input data is segmented online with the approach described in [151], and then incrementally clustered and represented in a tree structure [150]. Temporal sequences or relationships between the HMM motion primitives are captured in a motion graph and used to generate motion for a humanoid robot [149].

based on model similarity, and a group level HMM is learned for each subtree. Finally, a motion graph is learned to represent the sequential dependencies between the different motion primitives. Their approach has been tested with a 38 degree of freedom humanoid robot, with input from motion capture demonstrations using 34 markers locations on the human. The data consisted of 16 min of whole body motion from a single human subject, where the person did motions such as walking, squatting, kicking and raising arms. The video from this input data was manually segmented and labeled, providing ground truth which could be used for evaluation. The resulting system was able to achieve error rates below 10% on correctly identifying the motion primitives. The learned motion graph is successfully able to generate real-time continuous sequences of these motion primitives on a humanoid robot.

Similar to Kulic, the work of Kruger et al. [147] is focused on learning primitives from a continuous observation stream. In their work, each primitive is its own Parametric HMM which represents primitive actions and their effects on objects in a tabletop manipulation environment. In Lee et al. [156] a humanoid robot learns to imitate gestures (dance moves, free space gestures as well as contact gestures like high-five) from human demonstration, explicitly accounting for missing data in the demonstrations. The observation sequence includes all of the robot's joint angles, and, in the case of physical contact, additionally includes distance between the human hand and the robot [157].

Asfour et al. [18] introduce a method in which continuous motion trajectories are preprocessed to select "key points" when some feature of the trajectory reaches an extremum, changes direction, stops changing, or a minimum amount of time since the last key point has passed. This

---

[1]Image taken from: "Incremental learning of full body motion primitives and their sequencing through human motion observation," Dana Kulic, Christian Ott, Dongheui Lee, Junichi Ishikawa and Yoshihiko Nakamura, *The International Journal of Robotics Research*, Vol. 31, No. 3, pp. 330–345, © 2011 by the authors. Reprinted by permission of SAGE.

representation results in a sparse trajectory of keyframes as input to train an HMM for dual arm manipulation tasks. The HMM is used to detect temporal dependencies between the arms in dual-arm tasks.

The work of Calinon et al. [55], uses HMMs for skill learning, but combines this with a Gaussian Mixture Model (GMM) approach to address the different λ parameter estimation problems separately. They estimate the observation probability $P(Y|X)$ as a GMM, followed by using the Baum-Welch algorithm to estimate only the temporal sequencing $P(X_t|X_{t-1})$. The first step in this approach is k-means clustering of the demonstrations (using Bayesian Information Criterion to select the optimal number of clusters k). This clustering result is the initialization for an expectation-maximization (EM) algorithm to fit a multivariate Gaussian Mixture Model (GMM) with K components of dimensionality D:

$$p(y_i) = \sum_{k=1}^{K} \alpha_k \, p(y_i|k)$$

$$p(y_i|k) = \mathcal{N}(y_i; \mu_k, \Sigma_k).$$

The symbols $\alpha_k, \mu_k, \Sigma_k$ represent the prior, mean, and covariance matrix of the Gaussian distributions $k = 1...K$. As a final learning step the Baum-Welch algorithm is used to estimate just the transition probabilities between the K components of the GMM.

To reproduce a skill with this HMM/GMM model, Calinon et al. apply Gaussian Mixture Regression (GMR). This approach overcomes the problem that in order for an HMM to forward simulate a smooth trajectory in the observation space, in practice it needs to have a lot of training data. In the case of LfD we want to have good behavior after just a handful of demonstrations. For a D-dimensional variable y, and the means and covariance matrices already computed, the GMR regression is done along the time dimension, producing a canonical trajectory for the skill given the desired time-stamps, i.e., $P(y|t)$. A controller is then used to follow this trajectory. In [55], an optimal control formulation is used with inverse of the covariance matrices utilized as cost.

An alternative to the HMM/GMM combination is to explicitly add time as a dimension of the observation sequence [55]. In the combined approach, the GMM is doing the work of representing the spatial variance of the skill, while the HMM is representing the temporal variation seen in the skill. By adding time to the observation state, the temporal dynamics become part of the GMM as well. The state vector $y(t)$ is augmented with its time-stamp (or index), $\bar{y}(t) = [t; y(t)]$. But now it is important that all of the trajectories in the demonstration set are aligned temporally. To do this, the data is subsampled in the time dimension to a constant length, typically using Dynamic Time Warping. Now the GMM process is run (initialize with k-means and do EM to estimate parameters of the k components), and the resulting set of μ_k, Σ_k, and π_k where $k = 1 \ldots K$ is the model of the skill.

There are two main parameters for an approach utilizing GMM and GMR. One we have mentioned already is the number of Gaussians in the model. The second, not-so-obvious one is the relative weight of each dimension in the regression step. For example, consider a typical state-space for an end effector trajectory composed of time (1), position (3), and rotation (4) components, totaling 8 dimensions. However, time is measured in seconds, position is measured in inches and rotation is in a quaternion representation, i.e., they have different units. We would get differing results if we change the units, which effectively set the relative scale of each component. However, the most important relative scale is related with time, since we only use time as our independent variable in the GMR phase. In the source code accompanying [55], the unit of time is discarded and replaced with indices, effectively scaling the time dimension with the sampling rate.

The approach of modeling the skill with a GMM and reproducing it with GMR has been widely used since its introduction, e.g., [5, 14, 51, 52, 179]. Two drawbacks of the approach are its dependence on explicit time indices, and that it is essentially an open-loop controller. Khansari et al. introduced the Stable Estimator of Dynamical Systems (SEDS) approach [131], which bears similarities with the GMM+GMR method. The difference is that it directly learns a closed-loop policy and that it forces this policy to be stable. In [54], Calinon et al. extend the HMM/GMM+GMR approach, with an acceleration-based controller that is essentially a spring-damper system in which the HMM/GMM model of the demonstration trajectories acts as an attractor. Thus, the system is no longer open-loop but will track the desired GMR output and will force the system back to the known subspace in response to a perturbation. This is a very similar approach to that of DMPs discussed previously, the primary difference being that DMPs have a single demonstration as the attractor being tracked, whereas here the attractor is an entire model of the input demonstrations, which has the advantage of being able to encode several skill alternatives in the same model.

4.4   TECHNIQUES FOR HANDLING SUBOPTIMAL DEMONSTRATIONS

In all of the approaches mentioned in this chapter there is a big assumption that the input set of demonstrations is an accurate representation of the skill to be learned. There has been little to no testing of these approaches with demonstration trajectories of naïve end-users. One exception is [83], in which HMMs for welding tasks are learned via teleoperation input. Their evaluation included testing that three different users were able to successfully teach a variety of welding skills.

Despite the lack of testing with end-users, the notion of the potential sub-optimality of human demonstrations is widely recognized. When discussed in the literature, the most suggested approach is to try to identify and eliminate noisy or sub-optimal demonstrations, as seen in [64] and [232]. Even since 1995 this has been a topic of discussion; Kaiser et al. [126] provide an analysis suggesting five sources of sub-optimality with human teacher: unnecessary actions in the demonstration, incorrect actions, unmotivated action, demonstration scenario is too limited for generalization, or specification of a wrong intention. Their conclusion is a set of conditions under which demonstrations can be deemed sub-optimal and removed.

The removal of sub-optimal demonstrations is unsatisfying, since even a failed demonstration should provide some information about the skill. Grollman and Billard take such a view in their work, and argue that data from failed human demonstrations of a task should not be discarded [103]. Instead, the authors show that it is possible to build models from this data that can bias a robot's exploration to find a successful way to perform a novel task.

CHAPTER 5

Learning High-Level Tasks

In this chapter, we look at how the actions that were derived from motion trajectories in the previous chapter can be used to learn higher level tasks (Figure 5.1). While the line between high-level and low-level learning is not concrete, the distinction we make here is that techniques in this chapter assume the existence of a discrete set of action primitives that can be combined to perform a more complex behavior. As in the previous chapter, we begin by discussing the state space representation for this learning problem. We then organize policy learning approaches into three categories: learning a reactive task policy representing a functional mapping of states to actions, learning a task plan, and learning the task objectives. We go on to discuss the role that feature selection, reference frame identification and object affordances play in the learning process.

Figure 5.1: In this chapter we present techniques for learning high-level tasks.

5.1 STATE SPACES FOR HIGH-LEVEL LEARNING

Compared to the previous chapter, the algorithms in this chapter are targeted at learning more abstract high-level tasks. As a result, the demonstrations themselves are typically performed at a higher level. Instead of trajectories, the teacher's demonstrations consist of *action primitives*, such as *[pick up]*, selected from among a library of actions executable by the robot. Action primitives can be hand-coded, executed by a planner or learned through one of the techniques in the previous chapter. Their execution also typically lasts a non-negligible duration of time, and the actions themselves are often parametarized (e.g. *[pickup(targetObj)]*).

The state space is represented using a set of features, which describe the relevant state of the world and can take on continuous, categorical or binary values. Some algorithms use the changes observed in feature values to infer the pre- and post-conditions of action primitives. As discussed later in this chapter, the problem of identifying which features are relevant to a particular task is also an active research area.

Another powerful means of communicating the desired behavior to the robot is through the specification of task goals. Several algorithms enable the teacher to specify the goal state that needs to be achieved, such as a particular configuration of objects. Reward functions also capture the goals of a task. For example, consider a robot learning to navigate to a target location. We might assign a reward of +1 for reaching the goal, -1 each time the robot collides with an obstacle, and 0 otherwise. This type of model is referred to as a *sparse* reward function because it has a value of zero in most states, with a small number of non-zero values. Sparse reward functions are relatively easy to specify for most tasks. Unfortunately, they are difficult to learn from because the robot is given no feedback about its performance in most states. A *dense* reward function is one which contains a reward value for most states in the domain. In our navigation example, a dense reward function might calculate the value for a given state based on the distance to the nearest obstacle and the distance to the goal. Specifying a dense reward function by hand is a very challenging process that typically requires a significant amount of trial and error. Learning from Demonstration provides two options for generating this input through other means. In the first, the teacher can manually provide rewards to the robot while observing the robot's actions in a process called *shaping*, and in the second, demonstrations of the desired behavior by the teacher can be algorithmically converted to a reward function in a process called *inverse optimal control*.

5.2   LEARNING A MAPPING FUNCTION

The techniques presented in this section formulate LfD as a supervised learning, or function approximation, problem focused on learning a mapping from input states to output actions. In this representation, demonstration data typically consists of state-action pairs, or trajectories of state-action pairs, that are examples of completing the skill or task. Given these demonstrations, the goal of the algorithm is to reproduce the underlying teacher policy, which is unknown except for the (usually sparse and noisy) demonstrations, by generalizing over the set of available training examples. The result of learning is a *policy* model that outputs actions given states. The following texts provide an overview of supervised learning methods: [9, 146, 206]. In this book, we focus only on methods that have particularly strong applications for LfD.

Decision trees [210] have been shown to be successful in a number of LfD applications, and can be used both in the case of a few demonstration and many. For example, Sullivan et al. [243, 244] leverage decision trees as part of a hierarchical finite-state automaton learning framework. In this context, given a small number of demonstrations, decision trees with probabilistic leaf nodes are used to learn the probabilistic transition functions between automata states. In a very different setting, Crick et al. [75] crowdsource hundreds of demonstrations by allowing online users to teleoperate a robot through a maze using a web interface. Nearly 80,000 demonstration data points are then combined into a single policy learned by a decision tree, resulting in nearly perfect ability to replicate the maze task, as well as some degree of adaptation to other similar mazes. Interestingly, the authors report little benefit from pruning in this domain, hypothesizing that overfitting is not a major concern in their particular setting.

Memory-based techniques [4], sometimes known as lazy-learning algorithms, have also been applied to LfD. In the context of LfD, each learning instance represents a robot state, and the label corresponds to the action that was demonstrated in that state. For example, Saunders et al. [227] use k-nearest neighbors algorithm to learn a number of navigation tasks. Another example of instance-based learning is Case-Based Reasoning (CBR), which utilizes a library of past experiences (cases) to solve new problems by finding a similar past cases and reusing them in the new situation. CBR techniques can support more complex training examples than k-NN, including symbolic representations, and often rely on more complex similarity metrics. CBR has been successfully applied to autonomous robot control in a number of applications, such as indoor navigation [161] and autonomous robot soccer [220], but in these applications individual cases are hand-coded by the developers. Extensions to LfD, in which cases are generated directly from human demonstrations, have been shown in real time strategy games by Ontañón et al. [198], and in learning human-robot collaboration by DePalma et al. [38].

Bayesian methods have been used extensively in LfD. Rao et al. [214] present a human-inspired Bayesian model of imitation in which they frame the task learning problem as the computation of a set of actions that will lead to the goal state s_N, given a set of observed and memorized states $s_1, s_2, ..., s_N$. For example, consider a simple imitation learning task where the imitator has observed and memorized a sequence of states (for example, $S_7 \rightarrow S_1 \rightarrow ... \rightarrow S_{12}$). These states can also be regarded as the sequence of sub-goals that need to be achieved in order to reach the goal state S_{12}. The objective then is to pick the action that will maximize the probability of taking us from a current state $s_t = S_i$ to a memorized next state $s_{t+1} = S_j$, given that the goal state is $s_G = S_k$. In other words, we would like to select the action A_i that maximizes $P(a_t = A_i | s_t = S_i, s_{t+1} = S_j, s_G = S_k)$. Rao et al. show how structuring the learning problem within this probabilistic framework can enable the robot to infer the intent of the teacher and estimate the probability distribution over the goal states.

Finally, many supervised learning methods provide a measure of confidence in their classification and regression results. This can be very helpful in LfD to convey the degree of certainty of the robot's actions and regulate the robot's autonomy. Inamura et al. [118] and Lockerd and Breazeal [165] both use Bayesian methods to estimate the robot's classification confidence and communicate it to the teacher. Chernova and Veloso [66] present the Confidence-Based Autonomy algorithm, a supervised learning technique that leverages the classification confidence of the underlying supervised learning algorithm (the authors use Support vector machines and Gaussian mixture models) in order to regulate the robot's autonomy and prevent execution in low confidence areas of the state space. Grollman and Jenkins [102] utilize Locally Weighted Projection Regression to similarly estimate the robot's confidence with respect to selecting an action in the current state.

5.3   LEARNING A TASK PLAN

An alternative to presenting policies as a direct state-to-action mapping is to represent the desired robot behavior as a plan. Within the planning framework, the policy is represented as a sequence of actions that lead from the initial state to the final goal state. Actions are often defined in terms of *pre-conditions*, the state that must be established before the action can be performed, and *post-conditions*, the state resulting from the action's execution. Unlike other LfD approaches, planning techniques frequently rely not only on state-action demonstrations, but also on additional information in the form of *annotations* of goals or intentions from the teacher. Demonstration-based algorithms differ in how the rules associating pre- and post-conditions with actions are learned, and whether additional information is provided by the teacher.

Planning has been extensively studied in the software agents community, including in the context of LfD. Lent and Laird [260] present a method for learning non-deterministic plans based on demonstration traces annotated with goal transition data. Garland and Lesh [95] introduce an algorithm for learning a domain-specific hierarchical task model from demonstration. Within this approach, the teacher is able to annotate the sequence of demonstrated actions and provide high level instructions, for example, the fact that some actions can occur in any order. Note that in both of the above examples, state-action demonstrations are supplemented with additional information from the teacher to aid in generalization.

One of the earliest LfD works that was demonstrated on physical robots was Learning by Watching by Kuniyoshi et al. [152], in which a plan was learned for object manipulation based on observations of the teacher's hand movements. This, and other early work in this area [116], enabled robots to replay actions observed during demonstration, however, the learned models had little ability to generalize beyond the demonstrated environment. More recent work in this area has focused on generalizability, as well as techniques for learning complex plan structures through various interaction modalities. For example, Veeraraghavan and Veloso [262] present an algorithm for learning generalized plans that represent sequential tasks with repetitions. In this framework, a humanoid robot is taught the repetitive task of collecting colored balls into a box based on only two demonstrations. Nicolescu and Mataric [191] also contribute techniques for learning from multiple demonstrations, presenting a framework for learning behavior networks, a high-level task structure that models the interaction between abstract and primitive behaviors and their effects. The framework enables the robot to generalize across multiple demonstrations and to refine the learned model based on speech input from the teacher. Rybski et al. [225] incorporate speech to an even greater degree, enabling the user to generate complex plan models based on spoken dialog with the robot. The teacher presents the robot with a series of conditional statements which are processed into a plan, and the robot is additionally able to verify any unspecified parts of the plan by asking questions.

The above techniques construct plan representations based on demonstrations of discrete, high-level action primitives, such as *pick up(x)*. A number of techniques have also focused on bridging the gap between low-level trajectory input and high-level task learning, providing a means for extracting abstract task structures from motion trajectories. For example, Ehrenmann et al. [82, 90] introduce a system that records the movements of a human hand using a camera, data glove and a magnetic field based tracking system, segments the resulting data to identify trajectories and grasps, and then generates a hierarchical task representation based on the identified movements.

Niekum et al. [192, 193, 194] present a series of algorithms that draw from recent advances in Bayesian nonparametric statistics and control theory to automatically detect and leverage repeated structure in low level demonstrations in order to produce a task plan. At the core of the presented algorithms are Bayesian nonparametric models—models that do not have a fixed size, but instead infer an appropriate complexity in a fully Bayesian manner without overfitting the data or requiring model selection. These models are used to discover repeated structure in the demonstration data, identifying subgoals and primitive motions that best explain the demonstrations and that can be recognized across different demonstrations and tasks. This process converts noisy, continuous demonstrations into a simpler, coherent discrete representation. The resulting discrete representation can then be leveraged to find additional structure, such as appropriate coordinate frames for actions, task-level sequencing information, and higher-level skills that are semantically grounded. Finally, this collection of data is combined to construct robust controllers that use an understanding of the world to adaptively perform complex, multi-step tasks.

The authors demonstrate their approach by teaching a PR2 mobile manipulator to assemble a small table. The table consists of a tabletop with four pre-drilled holes and four legs that each have a screw protruding from one end. Eight kinesthetic demonstrations of the assembly task were provided, in which the tabletop and one leg were placed in front of the robot in various positions. In each demonstration, the robot was made to pick up the leg, insert the screw-end into the hole in the tabletop, switch arms to grasp the top of the leg, hold the tabletop in place, and screw in the leg until it is tight. An example of this progression is shown in Figure 5.2(a)-(f). To make the insertion more robust, a bit of human insight is applied by sliding the screw around the area of the hole until is slides in. This allows the insertion to be successful even when perception is slightly inaccurate.

The demonstrations were then segmented and used to generate a Finite-State Automaton (FSA), shown in Figure 5.2(g). Importantly, the authors note that initial demonstrations do not always cover all the possible contingencies that may arise during the execution of a task. In the case of table assembly, task replay was sometimes successful, but several types of errors occurred intermittently. Two particular types of errors that occurred were (a) when the table leg was at certain angles, the robot was prone to missing the grasp, and (b) when the leg was too far from the robot, it could not reach far enough to grasp the leg at the desired point near the center of mass.

To address this problem, Niekum et al. use interactive corrections, which provide additional training data to the model. These corrections are provided by the user at the time of failure and are treated as additional demonstrations that can be segmented, used to improve the structure of the

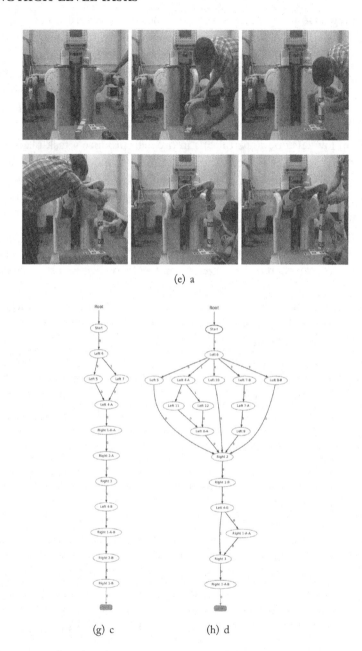

(e) a

(g) c   (h) d

Figure 5.2: From [194]: (a)–(f) A kinesthetic demonstration of the table assembly task. (g) FSA structure for the table assembly task from original demonstration data. (h) FSA structure with the addition of data acquired through interactive correction.

FSA, and provide additional examples of relevant primitives. Together, this allows for iterative, incremental learning and improvement of a complex task from unsegmented demonstrations. For table assembly, interactive corrections were provided to recover from the above listed contingencies. In the first case, a re-grasp was demonstrated, and then the task was continued as usual. In the second case, the robot was shown how to grasp the leg at a closer point, pull it towards itself, and then re-grasp it at the desired location. After the interactive corrections were collected, the old data was re-segmented with the two new corrections and used to re-build the FSA. Figure 5.2(h) shows the FSA that results from this process. Using this new FSA, the robot was able to recover from both types of errors in novel situations.

5.4 LEARNING TASK OBJECTIVES

An alternative to representing a task as a set of actions, is for the robot to infer the purpose behind the observed behavior and model the task in term of its goals or objective. Modeling goals has been explored in a number of programming by demonstration techniques [94, 260].

The work of Breazeal et al. [36], inspired by the goal-oriented nature of human learning, takes this as the primary task in an LfD process. In this work, the robot is instructed through speech commands on how to achieve a new task. The learning algorithm pays attention to what actions the robot is asked to perform and infers goals for these actions by comparing the perceptual state before and after the action. The algorithm expands a hypothesis space about this state-change goal of all representations consistent with the current task example, i.e., expanding a version space of the goal concepts consistent with the demonstration [41]. The result is a lattice of hypotheses consistent with the positive examples, ordered from most specific to most general. Hypotheses are eliminated as more examples of the concept are seen. The current best hypothesis is used for task execution, and is selected by Bayesian likelihood, where the likelihood of each of the hypotheses is calculated according to $P(h|D) \propto P(D|h)P(h)$. The data, D, is the set of all examples seen for the task. $P(D|h)$ is the percentage of the examples where the state change seen is consistent with the goal representation in h. For priors, $P(h)$, the algorithm prefers a more specific hypothesis over a more general one. For example, when a task is first learned, every hypothesis is equally represented in the data, so the algorithm chooses the most specific representation for the next execution. This same formulation of goal learning has been extended by Gray and Berlin to the context of visual perspective taking [39], and by Chao et al. for the purpose of perceptual symbol grounding [61].

In the context of Reinforcement Learning, the goal of the task is captured in the representation of the reward function. For example, the learner may receive a high reward for reaching the target state or for maintaining a trajectory free of collisions. Thus, approaches that *learn* the reward function from human demonstrations can be considered a form of goal learning. One of the earliest works in which the reward was inferred from demonstration was by Atkeson and Schaal [19], in which the robot infers the task model and reward function from demonstrations, and then uses this information to derive a task policy for a pole-balancing task. In this algorithm, a specific reward function is not derived, instead greater reward is given for states similar to those demonstrated by a human. In a similarly motivated approach, Guenter and Billard [105] present an algorithm for learning motion trajectories in which greater reward is given to the robot for similarity between the executed trajectory and the demonstration, and for achieving a position close to the goal. Using this representation, the authors present an algorithm that enables the robot to adapt to changes in the task, such as a new goal location or an obstacle in the robot's movement path, through the use of exploration and RL.

Konidaris et al. [144] incorporates goals in a different way by leveraging the options framework, a hierarchical RL formalism [246]. The authors present CST, an online algorithm for constructing skill trees from demonstration trajectories. CST segments a demonstration trajectory into a chain of component skills, where each skill has a goal and is assigned a suitable abstraction from an abstraction library. These properties permit skills to be improved efficiently using a policy learning algorithm. Chains from multiple demonstration trajectories are merged into a skill tree.

In the above techniques, human demonstration influences the reward function, but is not used to derive the reward function explicitly. The problem of deriving an explicit reward function from demonstration data is referred to as *Inverse Reinforcement Learning (IRL)* [189]. In this paradigm, demonstrations observed by the learner are used to generate the reward function, which is then used by the learner to construct its own policy and solve the target task. Abbeel and Ng pioneered the use of IRL techniques in robotic application, demonstrating the use of this technique first in simulation [3] and then in teaching autonomous helicopter flight [1, 2]. Their work was extended in Kolter et al. [143] to consider the decomposition of task demonstration into hierarchies. Syed et al. [247, 248] have explored this problem from a game-theoretic perspective, and proposed algorithms to learn from demonstration with provable guarantees on the performance of the learner.

The work of Ramachandran and Amir [213] introduced Bayesian Inverse Reinforcement Learning, in which IRL is cast as a Bayesian inference problem. Given a prior distribution over possible target tasks, the algorithm uses the demonstration by the expert as evidence to compute the posterior distribution over tasks and identify the target task. Lopes et al. [167] and Babes et al. [21] improved on the computational complexity of this method by taking advantage of the underlying IRL problem structure and gradient-based methods to determine the maximum likelihood task representation. In two other closely related techniques, Ziebart et al. [272] and Neu and Szepesvari [188] also present gradient based IRL methods.

Finally, a series of papers by Ratliff et al. [217] and Silver et al. [232] present the LEARCH (LEArning to seaRCH) algorithm that uses demonstrations to infer the objective function that is used to score potential future actions and derive optimal plans. For example, in an offroad autonomous navigation domain, the user is able to draw a path on a map representing the desired behavior, such as preferred terrain type. Maximum Margin Planning is then used to derive a cost function that assigns values to states in a way that makes the demonstrated trajectory optimal. This cost function can then be used with a wide variety of path planning algorithms, such as

A*, to plan trajectories that imitate the teacher's preferences in new locations. One limitation of LEARCH is that by trying to imitate the teacher, the algorithm is limited by the quality of the teacher's demonstrations. Zucker and Bagnell [273] present an extension based on Reinforcement Planning, which propagates the reward signal back through the controller and planner to the parameters of the underlying cost function, this allowing the cost function to be improved beyond the original teacher demonstration. In similarly motivated work, Ollis et al. [196] present a Bayesian approach for calculating the terrain costs based on demonstrations.

5.5   LEARNING TASK FEATURES

Correctly interpreting demonstrations is critical to the success of the above learning techniques, which brings us to the problem of feature selection. Feature selection is the process of selecting a subset of relevant features for use in model construction through elimination of redundant or irrelevant features [106, 164]. Within LfD, the problem of feature selection has typically been ignored under the assumption that the teacher or programmer has made the best selection of features manually a priori. However, it is important to remember that in many target applications the LfD end user is not expected to have sufficient technical knowledge to perform feature selection accurately.

Cobo et al. [71] present *Abstraction from Demonstration (AfD)*, an algorithm that learns a policy for an MDP by building an abstract space S^α and using RL to find an optimal policy that can be represented in S^α. AfD obtains S^α by selecting a subset of features from the original state space S with which it can predict the action that a human teacher has taken in the set of demonstrations. Learning in S^α can be significantly more efficient than learning with the full feature set because a linear reduction in the features leads to an exponential reduction in the size of the state space. Additionally, the authors show that this use of naive human demonstrations for selecting the feature space for RL is significantly more sample efficient than using the demonstrations for supervised learning. Levine et al. propose an algorithm that selects relevant features to represent the reward function in Inverse Reinforcement Learning by building logical conjunctions of the features to the example policy [158].

In an idea that is complementary to feature selection, Meriçli et al. [176] contribute the Multi-Resolution Task Execution algorithm. The authors present a general framework that employs a set of detail resolutions, in which each resolution has its own state and action representations, and an algorithm using these representations to perform the task. Over the course of a training session, a teacher observes the robot executing the task using hand-coded algorithms, and intervenes if the current algorithm needs a correction, or if the detail resolution in use is too coarse to cope with the current situation. The robot learns a detail switching policy for deciding which detail resolution to use in a particular state while also building up individual corrective demonstration databases for the algorithms at each detail resolution. During the autonomous execution of the task, the robot first chooses the most convenient detail resolution to run at, and then computes the action to be performed in the perceived state at the selected detail resolution.

Dong and Williams [84, 85] consider the problem of autonomously identifying what features or relations, if any, are characteristic of a particular demonstrated motion. In this context, the relevant motion variables are those that are preserved over different demonstrated trials of that motion, while other motion variables may vary due to changes in the environment or the human's movement. For example, demonstrated sequences of the motion "move box to bin" will show a pattern whereby the robot end effector starts at the location of the box, makes contact with it, moves to the location of the bin, and breaks contact with the box. The system will learn that the distance between the robot effector and the box is a relevant motion variable at the beginning of the motion, and that the distance between the robot effector and the bin is a relevant motion variable at the end of the motion. The system will also learn that the positions of any other objects known in the environment are not relevant to this motion.

5.6   LEARNING FRAME OF REFERENCE

Related to the problem of feature selection is the problem of identifying the frame of reference. Most physical actions are performed within a particular reference frame, and correctly inferring that reference frame can help to understand the actions, their goals, and to generalize over observed behavior. For example, the trajectory an arm follows while reaching towards several different objects may be different, but significant similarities can be identified when the arm motion is considered relative to the target object. Ideally, such analysis would enable the robot to identify all of the above motions as belonging to the same class of behaviors.

Cederborg et al. [58] consider three possible reference frames in which a motion can be performed: relative to the starting position, relative to the robot frame, and relative to an object position. The desired velocity estimated at every state is the weighted sum of the desired velocities in each reference frame, where the weights reflect how well the data trains in each frame. One limitation of this representation is that each motion is restricted to exactly one of the reference frames. As a result, it is not possible to model actions that span multiple reference frames within a single model (e.g. picking up and moving an object to an absolute location). Niekum et al. [193] similarly consider a single reference frame for each motion, determining the relevant reference frame by clustering the end point of each motion within the candidate coordinate frames.

Dong and Williams [84, 85] present a more general model that enables different reference frames to be applied to the beginning and end of the motion. Demonstrated motion trajectories are segmented at a subset of time steps determined by the operator, and the learning algorithm determines relevant motion variables at the endpoints of these segments, as discussed in the previous section. Typically, these are time points corresponding to a qualitative change in the behavior of the task, such as the robot making or breaking contact with an object. For each of the continuous, discrete, position, and orientation input variables of a demonstrated motion, the algorithm considers up to five possible modes for candidate motion variables: absolute start, absolute end, relative to initial, relative to object at start and relative to object at end (Figure 5.3). Cluster-

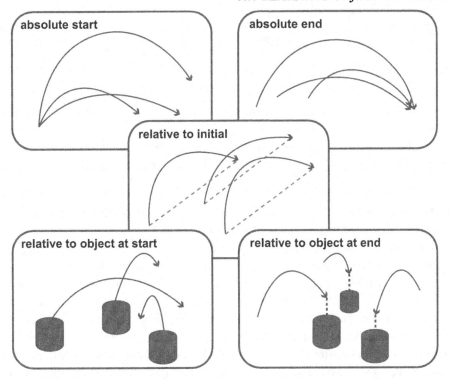

Figure 5.3: Example illustration of five possible ways that motion variables can be relevant. Arrows refer to the robot end effector trajectories [85].

ing is then applied to determine the correct reference frame for the beginning and end of each trajectory.

5.7   LEARNING OBJECT AFFORDANCES

Another line of work that can be put into the realm of high-level task knowledge is affordance learning. Introduced as a concept by Gibson [96], affordances are properties of the environment that afford a certain action to be performed by a human or an animal. The goal of affordance learning is to build a model of the relation between objects in the world and the robot's action repertoire. Identifying object affordances enables the user to categorize objects by their function, and thus this representation has been utilized in robotics as a compact and useful representation for manipulation skills (e.g., Figure 5.4).

In many works the robot learns affordances from its own exploration by acting upon objects in the environment and observing the reaction. In [92], a robot pushes, pokes, pulls, and grasps objects with its end-effector, thereby learning about rolling, sliding, etc. One of the benefits of

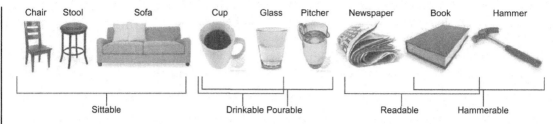

Figure 5.4: Representing objects in terms of functionality and affordances. Top: Semantic, appearance-based categories. Bottom: Functional, affordance-based categories [133].

learning through this type of interaction is that no correspondence problem is present, enabling the robot to directly observe the effects of its own actions.

Affordances can also be learned from demonstrations by visually observing a human or other robot making use of object affordances. Thomaz and Cakmak [47] compare learning of object affordances in two settings, a social interaction that is structured by a human partner, and a non-social setting in which the robot is presented with a systematic set of object configurations to explore. The authors make six observations about how the social data set differs from the non-social data set: people have a more balanced set of positive and negative examples; they intuitively structure the environment with respect to complexity, both in number of examples per object and order of examples; social data sets have a greater representation of rare affordances; and people's actions in the workspace can be used to infer action goals. Ultimately, the affordance classifiers trained in the social setting were more effective at predicting rare affordances since people focused on these, and they performed on par with non-social SVMs on the more common affordances.

Learning from observation of human actions has also be addressed in several other works. In a series of papers by Lopes et al. [168, 169] and Montesano et al. [177], the authors present techniques for affordance-based imitation learning for manipulation tasks. Affordances in this framework are represented as Bayesian networks that encode the dependencies between actions, object features and the effects of those actions. The learning of affordances with a Bayesian network is performed in two phases. First the structure is learned using Markov Chain Monte Carlo, and then the parameters of each node are estimated in a separate step. The resulting model can then be used to interpret the effects of observed actions, as well as to predict the effects of the robot's own actions in terms of the world state. Alternate methods for learning objects affordances for manipulation have also been presented. Aksoy et al. [6] introduce a representation of the relations between objects at decisive time points during a manipulation, providing the ability to encode the essential changes in a visual scenery in a condensed way such that a robot can recognize and learn a manipulation without prior object knowledge. The presented algorithm continuously tracks image segments in the video, constructs a dynamic graph sequence and stores topological graph transitions within a matrix structure called a semantic event chain. Objects with similar manipulation properties can then be recognized through sub-string search algorithms. Kjellström et al. [133] present an algorithm based on Conditional Random Fields [154] that enables manipulated objects and human manipulation actions to be categorized in the context of each other by simultaneously identifying and classifying human hand actions and the objects involved in the action.

In [263], Veloso et al. take a different approach to affordance learning, focusing on full-body environmental interactions instead of simply object manipulation. The authors present the FOCUS algorithm, which models inanimate objects in the environment by structural and functional definitions. The structural part of the model aims at capturing a simple and generalized visual definition of an object through robust feature detectors. The functional part of the model captures the affordance properties of that object, such as the fact that chairs are for sitting. Objects in the environment are recognized by associating an observed action with a particular environmental feature. The classification of the object is dependent upon the specific activity for which it is used by the person. As an example, if a robot equipped with the FOCUS algorithm observes a human walk through a room and sit in a chair, then the visual features nearest to where the human sat would fall under the classification of chair. In this case, a chair is anything that a human will sit upon. This classification of chair could very well be given to a small table, a couch, or even a heat register if the human chose to sit upon it. The interesting aspect of this functional view is that it can be rather robust to the specific environment conditions of the signal capture. By connecting sitting with non-ambitious definition of a chair, the problem is converted mainly into motion recognition and the robustness to the environment is achieved. By finding one object in the image, we can then generalize and find multiple similar objects.

5.8   TECHNIQUES FOR HANDLING SUBOPTIMAL DEMONSTRATIONS

Demonstration errors made by the user can be classified into one of three categories: (1) correct but suboptimal (e.g. containing extra steps); (2) conflicting/inconsistent (e.g. sometimes the user demonstrates to go left and sometimes right from the same state); and (3) entirely wrong (e.g., incorrect action selected). Techniques for improving on suboptimal demonstrations are the focus of Chapter 6, where we present a variety of methods for improving the policy based on corrections by the teacher or independent exploration by the robot. Conflicting and incorrect demonstrations (categories 2 and 3) are often grouped together in the literature and simply referred to as "noisy demonstration data." However, some research suggests that making the distinction between these categorizations can help to understand the cause of the noise, leading to improved learning methods.

In many domains, robots encounter *equivalent action choices*—situations in which multiple actions are equivalently applicable. For example, a moving robot that encounters an obstacle directly in its path has the option of moving left or right to avoid it. If the space is empty, both directions are equally valid for performing the desired task. Similar choices can arise in many other situations, such as deciding among objects of equal value. Human demonstrators faced with a choice of equivalent actions typically do not perform demonstrations consistently, instead selecting among the applicable actions arbitrarily each time the choice is encountered. As a result, training data obtained by the robot is conflicting, such that identical, or nearly identical, states are associated with different actions. Distinguishing conflicting demonstrations from random (or systematic) teacher error can enable the robot to correctly interpret the meaning of the demonstrations—mainly that in this particular situations multiple actions are applicable. Chernova and Veloso [67] present an algorithm for identifying regions of the state space in which data from multiple classes overlaps as a result of inconsistent demonstrations. For these regions, the authors make the assumption that a valid robot action can be selected at random among the conflicting data classes. The choice between multiple actions is then modeled explicitly within the robot's action policy through Option Classes. This automated approach does not require the teacher to predefine or demonstrate special choice actions, extending instead from the person's natural demonstration technique. A similarly motivated approach is also introduced by Butterfield et al. [43] for real-valued demonstrations, in which the authors use a multi-valued function regressor designed for time-series data to discover latent variables which represent hidden objectives in the demonstrated data.

Techniques for handling simply noisy demonstrations are far more abundant, although most approaches rely on the properties of the underlying machine learning technique for simplicity. For example, an LfD technique that uses decision trees to learn the task policy inherently adopts the decision tree's ability to robustly handle noisy data. If sufficient demonstration data is collected, statistical methods can also be used to filter out outliers prior to learning. An example of this type of approach is presented by Breazeal et al. [38], in which data from hundreds of teachers is analyzed to identify dominant patterns, which are then used to filter out noisy actions.

5.9   DISCUSSION AND OPEN CHALLENGES

As can be seen from the previous sections, task-level LfD has been explored not only through a broad range of Machine Learning techniques, but also in the context of challenges such as feature selection, goal extraction and affordance learning. In this section, we discuss a number of open challenges that must be addressed in order to promote advances in this research area. We organize our discussion around two themes: algorithmic advances and usability.

From the algorithmic perspective, many parallel advances are being made in learning from different forms of human input, but most of the existing techniques in these areas have been verified in single domains and in isolation. Integration of these ideas into a unified learning model and more extensive testing with varied domains and real users is needed to gain understanding of how learning can be scaled up to more complex domains. Among current methods, two areas that are rarely addressed are techniques for learning tasks of increasing complexity based on skills previously learned with LfD, as well as the use and learning of parametrized actions (e.g., *pickup(x)* instead of separate *pickupItem1* and *pickupItem2* actions), which is critical for effective scalability in learning.

In conjunction with algorithmic development, greater focus needs to be placed on advancing the usability of LfD techniques. Human-robot interaction plays a central role in this research area, and is critical to eventual successful deployment of LfD techniques in real world applications. From this perspective, many of the manual processes currently performed by developers will need to become either automated or accessible to the target user. This includes processes such as the selection of relevant features and parameter tuning for the underlying algorithm. Finally, it is critical that the user be able to understand what the robot has learned and its degree of proficiency at the task. Toward this end, more research needs to be conducted on transparency techniques and ways to communicate the extent of the robot's knowledge to the user.

CHAPTER 6

Refining a Learned Task

The teaching and learning processes of a situated learning interaction are tightly coupled, and a good instructor is able to maintain a mental model of the learner (e.g., what is understood, what remains unknown) in order to provide appropriate scaffolding to support the learner's needs. Examples of scaffolding mechanisms, introduced in Chapter 2, include attention direction, feedback, regulating the complexity of information, and guiding the learner's exploration. In general, this is a complex process where the teacher dynamically adjusts their support based on the learner's demonstrated skill level. The learner, in turn, helps the instructor by making their learning process transparent through communicative acts, and by demonstrating their current knowledge and mastery of the task.

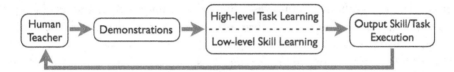

Figure 6.1: This chapter addresses several ways in which a learned model can be refined and improved interactively with the human teacher.

In this chapter, we discuss techniques for achieving tightly coupled interaction in LfD. Having covered the individual pipeline components, we address the iterative nature of the LfD process (Figure 6.1). First, we consider the difference between incremental versus batch LfD with a human teacher. Then we cover three different approaches to refinement: extensions to the RL algorithm that incorporate human input, human-initiated corrective demonstrations and robot-initiated Active Learning (AL) methods.

It is important to note that policy refinement does not require a human teacher. Refinement of a policy (whether learned through demonstration or some other means) can also occur through exploration and practice. A recent survey of the use of RL in robotics provides a summary of such techniques [140]. As demonstrated by Pastor et al. [202], such methods can be particularly powerful when training a task that humans find challenging themselves. In the sections below, however, we focus our discussion specifically on methods that seek to further speed up learning by integrating human input into the refinement process.

6.1   BATCH VS. INCREMENTAL LEARNING

We first consider the impact of performing learning as a batch versus an incremental learning process. Within LfD, *batch learning* resembles much of traditional Machine Learning in which the process of acquiring a dataset is independent of using the dataset to learn a model. In this paradigm, the teacher demonstrates the task one or more times in order to record a dataset of examples, and this data is then used to generate the learned policy. While interaction between the teacher and robot does occur during the demonstration process, it is loosely coupled. *Incremental learning* techniques interleave demonstrations and task execution, alternating between teacher control (demonstrations) and autonomous control by the robot in a tightly coupled interaction. The technique for determining how and when the switch of control is regulated lies at the heart of such methods. Proposed methods have included enabling the teacher to observe autonomous behavior and provide corrections for mistakes [14, 66, 102, 271], as well as querying mechanisms that enable the robot to halt execution and request help [45, 66].

Given the choice of available batch and incremental learning algorithms, it is important to consider the impact that a tightly vs. loosely coupled interaction has on the learning process. An experiment by Zang et al. asked this question, and found that simply seeing an agent's behavior while it is learning significantly improves a human's teaching demonstrations [271]. We detail their experimental findings here as motivation for the need to consider refinement in an LfD interaction.

The authors compared two teaching paradigms, interactive LfD and batch LfD, in an experiment with a scaled down Pac-Man game (7x8 grid). The human teacher provides demonstrations by playing the game, and the agent performs Q-Learning along the demonstrated trajectory to build a policy. Half of the teachers in their experiment taught in batch mode, demonstrating 30 consecutive games. After the 30 demonstrations, the agent played 60 additional games to learn on its own. Thus, the final policy is a result of learning from these 90 trajectories. The other half of the teachers taught in interactive mode. They first demonstrate 15 games, then the agent learned on its own for another 30 games (this process was performed opaquely, with all animations showing agent actions turned off). After this initial learning period, players were then allowed to see the agent as it learned and explored on its own (i.e., they were able to watch the agent controlled Pac-Man move around on the board). This continued for the remaining 45 games of the experiment. Players were asked to watch agent play, and if they deemed necessary, provide corrective demonstrations. To provide a corrective demonstration, they pause the game and rewind play to an appropriate point from which to provide a demonstration of what Pac-Man should have done. Players were allowed a maximum of 15 corrective demonstrations, thus the batch and interactive modes both end up with 30 demonstrations from the teacher.

In analyzing the averaged learning curves for each of these two groups, Zang et al. report that initially, batch and interactive performances were comparable. But around game 60, the interactive group started outperforming batch with a statistically significant difference between the groups, with that difference becoming more pronounced by the end of all 90 games.

Next, Zang et al. show evidence that interactivity improves the teacher's mental model of the learner and encourages them to change their teaching strategy (i.e., give better demonstrations) based on this. In exit interviews, 70 percent of interactive participants said they changed their teaching strategy during the session. Empirically this can be seen by contrasting the teacher's demonstrations from the first fifteen games before they have a chance to see learner performance with those used in the remaining teaching demonstrations. Looking at the policy difference between these, i.e., the shift in teaching strategy, shows a statistically significant difference for those participants who said they changed policies over those who said they did not. In other words, participants who said they changed policies actually did.

Moreover, the change was based (at least in part) on learner performance. Relative initial performance of the agent was shown to predict the amount of policy change. And it was this strategy change (at around game 60) that produced the jump in learning performance for the interactive group. The adapted strategies accelerated the learning rate by causing low performing teachers to give better demonstrations making their performance similar in the end to teachers that gave good demonstrations from the beginning.

These results provide concrete data to help guide practitioners on how best to obtain demonstrations from non-expert humans for LfD applications. It highlights the importance of the teacher having an accurate assessments and understanding of the learner in order to give good demonstrations.

6.2   REINFORCEMENT LEARNING BASED METHODS

One of the most common methods for refining a learned policy is through Reinforcement Learning. As noted earlier, RL can be used to refine an existing policy without human involvement by learning directly from environmental reward and exploration. Details of different Reinforcement Learning techniques can be found in surveys of the field [125, 140, 245]. In this section, we focus our discussion on techniques that integrate human input into the refinement process in an attempt to further speed up learning over autonomous exploration methods.

In traditional RL, the reward function used to refine the policy through RL is hand-coded by the developer. However, determining the reward values is a challenging process that requires a lot of trial and error. There are two ways to instead obtain the reward function from human input. The first, is to derive the reward function directly from the demonstrations through inverse optimal control methods. These techniques were described in Section 5.4.

The second method for integrating user input is a technique called *interactive shaping*, which is defined as interactively training the learning agent by providing positive and negative reward signals directly from the human. This technique assumes that a human trainer, who knows the goals of the task, observes the robot's actions and provides rewards that are positively correlated with the trainer's assessment of recent state-action pairs.

There is an extensive body of work on shaping [33, 59, 120, 129, 134, 208, 226, 250]. Here, we will explore this research area by studying the TAMER framework, shown in Figure 6.2, which

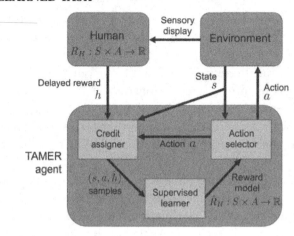

Figure 6.2: Framework for Training an Agent Manually via Evaluative Reinforcement (TAMER) [138].

was developed by Knox and Stone [134, 135, 138]. Within TAMER, the environmental reward function R is removed from the task model, resulting in an $MDP \setminus R$ representation. Instead, all reward comes from the human trainer, which is then modeled by the human reward function $\hat{R}_H$. Given the current state description, the agent's goal is to choose the action that will receive the most reward from the human. To do this, the agent uses $\hat{R}_H$ to greedily choose the action that is expected to earn the most reward. After learning an accurate model of the human's reward, the agent can continue to perform the task in the absence of the human, choosing actions that are predicted to maximize the received reward if the human were present. Note that the agent tries to maximize immediate reward, not expected return (i.e. a discounted sum of all future reward), under the assumption that the human trainer is already taking each action's long-term implications into account when providing feedback.

While the TAMER framework was originally applied in virtual worlds, recent work has demonstrated its use for training physical robots. Figure 6.3 shows the results of a study in which a mobile robot is taught to perform five interactive navigation behaviors: go to, keep conversational distance, look away, toy tantrum and magnetic control. On the left, the figure shows iconic illustrations of the five behaviors performed relative to a training artifact located in the environment. On the right, are heat maps showing the reward model that was learned at the end of each successful training session. The robot is shown as a transparent birds-eye rendering in each image, positioned to face the top of the page. The map colors communicate the value of the reward prediction for taking that action when the artifact is in the corresponding location relative to the robot.

In this particular implementation of TAMER, the human reward function $\hat{R}_H$ is modeled by the k-nearest neighbors algorithm, with a separate sub-model per action that estimates

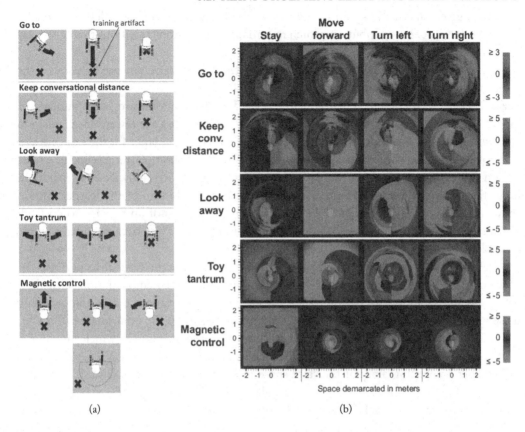

(a) (b)

Figure 6.3: (a) Iconic illustrations of the five interactive navigational behaviors. Each gray square represents a category of state space. The arrow indicates the desired action in such state; lack of an arrow corresponds to the stay action. (b) Heat maps showing the reward model that was learned at the end of each successful training session. A legend indicating the mapping between colors and prediction values for each behavior is given on the right.

the expected reward for performing the given action in the current state. During autonomous execution, the robot selects the action with the highest expected reward. A key insight of the TAMER framework is that the problem of credit assignment inherent in reinforcement learning is no longer present with an attentive human trainer. The trainer can evaluate an action or short sequence of actions, consider the long-term effects of each, and deliver positive or negative feedback within a small temporal window after the behavior. Assigning credit within that window presents a unique challenge, in many domains the frequency of time steps is too high for human trainers to respond to specific state-action pairs before the next one occurs. The Credit Assigner module within TAMER addresses this problem by modeling the expected delay in receiving feedback for a given state-action pair. In the case of the robot domain above, delay is modeled as a Uniform(-0.8 s, -0.2 s) distribution, which rewards all state-action transitions that fall within that temporal range.

Human input can also be used to modify the action selection mechanism of the Reinforcement Learning algorithm itself [69, 136, 137, 148, 170]. For example, Thomaz and Breazeal [252, 253] combine environmental reward with human reward and a human *guidance* input. A key insight of this work is that human teachers often use the reward mechanism to provide *anticipatory reward*, intended to represent future directed guidance for the learning agent, in addition to providing the feedback for past actions. Reinforcement Learning is unable to interpret reward as an anticipatory signal. As a result, the authors introduce guidance as a separate human input that enables the human teacher to direct the action selection of the learner. Their result show that this combined approach provides performances gains of up to 50% over a feedback-only method. Originally demonstrated in simulation, this work was later extended to a physical robot domain [242].

Another way to formulate LfD as an RL problem is for the teacher to perform demonstrations of the correct actions instead of providing reward. These demonstrations provide a good baseline policy from which traditional RL can take over and improve on. A unique strength of these techniques is that RL enables the learning robot to *surpass* the performance of the teacher, which is particularly helpful when dealing with novice users or tasks in which human demonstration is suboptimal. Following this principle, Smart and Kaelbling [236] use demonstration to highlight interesting areas of the state space in domains with sparse rewards. Teleoperation is used to show the robot the reward states, eliminating the long periods of initial exploration in which no reward feedback is acquired. Conn and Peters [74] similarly explore the use of RL to learn an optimal path for navigation, using demonstrations acquired from a supervisor to bias action selection.

Integrating elements of both reward and demonstration, the supervised actor-critic reinforcement learning algorithm [221] enables the teacher to influence the performance of the learner through both reward and action demonstrations. Using reward feedback, the teacher is able to penalize negative and reinforce positive behavior, and action demonstrations are used to highlight recommended actions and to suggest promising directions for state exploration. This technique has been applied to a basic assembly task using a robotic manipulator.

As seen in the examples above, human input can be integrated into Reinforcement Learning in a variety of ways, ranging from demonstrations to reward feedback. This raises an important question for this research area: What is the right level of human involvement? In addressing this, it is useful to characterize the level of human interaction as a spectrum from shaping to exploration. On the shaping end is a system that is completely dependent on a human instruction, and on the exploration end is a system that learns through self exploration with limited input from a human partner. In prior works that introduce a human to a machine learning process, we see a dichotomy with respect to this spectrum with several approaches at either extreme. Additionally, the level of human interaction generally remains constant throughout the learning task. A challenge for future work is to break this dichotomy and build interactive learning systems that can effectively operate along the entire shaping-exploration spectrum, successfully incorporating both self and social learning strategies within a single framework.

Finally, while our examples above have focused largely on high level learning domains, RL has also been widely applied to refining low level motion trajectories. The survey of Reinforcement Learning in Robotics by Kober et al. [139, 140] provides a detailed discussion of these methods.

6.3   CORRECTIVE REFINEMENT FROM THE TEACHER

In this section we examine techniques for policy refinement that rely on corrective demonstrations. In this context, we assume that the policy refinement process is controlled by the teacher; the teacher observes robot task performance and provides corrections to identified mistakes. Letting the teacher guide and control the policy refinement process is the most common way to improve the performance of an existing policy. Techniques for controlling the flow of information include placing the robot in a particular (often new) situation, providing more execution examples, and providing critique. These methods are in contrast to learner-guided refinement techniques, such as Active Learning, which are discussed in the next section.

Corrections typically indicate a preferred state or action, and within the existing literature corrections are most frequently provided within action spaces where the actions are discrete and of significant time duration, and therefore allowing the user time to evaluate and respond to each action choice. Nicolescu and Mataric [191] present a learning framework based on demonstration, generalization and teacher feedback, in which training is performed by having the robot follow a human teacher and observe the teacher's actions. A high-level task representation is then constructed by analyzing the experience with respect to the robot's underlying capabilities. The correct action from a discrete set is provided by a human teacher. Lockerd and Breazeal [36, 165] demonstrate a robotic system where high-level tasks are taught through social interaction. In this framework, the teacher interacts with the robot through speech and visual inputs, and the learning robot expresses its internal state through emotive cues such as facial and body expressions to help guide the teaching process. The outcome of the learning is a goal-oriented hierarchical task model. Meriçli et al. [175] also enable the teacher to correct a task model through dialog. Chernova and Veloso [66] introduce a mixed-initiative learning algorithm, Confidence-Based Autonomy (CBA), which allows the teacher to additionally perform corrective demonstrations when an incorrect action is selected by the robot. Inamura et al. [118] present a similarly motivated method based on Bayesian Networks.

Correction methods for continuous trajectories have also been studied. Argall, Browning and Veloso introduce a series of algorithms that build policies through a combination of demonstration and teacher feedback [15]. The algorithms *Binary Critiquing* and *Advice-Operator Policy Improvement* use binary performance flags and corrective advice as feedback, respectively, to refine motion control policies learned from demonstration. The algorithm *Feedback for Policy Scaffold-*

ing uses multiple types of feedback to refine primitive motion policies learned from demonstration, and to scaffold them into complex behaviors. The algorithm *Demonstration Weight Learning* treats different feedback types as distinct data sources, and through a performance-based weighting scheme, combines data sources into a single policy able to accomplish a complex task. The authors demonstrate the use of each of these four algorithms for robot navigation.

Finally, in providing feedback to the learner, the teacher must decide not only what information to provide, but also *when*, or for what states. Several approaches have explored techniques for communicating the robot's confidence in its task execution to the user in order to aid the user in identifying problem areas. For example, Grollman and Jenkins present the Dogged Learning algorithm [102], a confidence-based learning approach for teaching low-level robotic skills. In this algorithm, the robot indicates to the teacher its certainty in performing various elements of the task. The teacher may then choose to provide additional demonstrations based on this feedback.

6.4   ACTIVE LEARNING

One common assumption held by many demonstration learning approaches is that the teacher, as an expert at the task, always knows when and which demonstrations are required to improve robot performance based on observations of the robot's behavior. However, the robot and teacher represent knowledge in different ways. As a result, the teacher does not always know what additional information the robot requires. In some cases, the teacher is likely to perform redundant demonstrations that provide little information for improving the policy [66].

In this section, we consider the robot as a more active participant that provides feedback to the teacher about the learning process, indicating uncertainty or even asking questions about the task. This approach forms a natural extension of demonstration learning as it builds upon an existing interaction framework, while allowing the robot to help guide the learning process.

Within machine learning research, Active Learning [73] enables a learner to query an expert and obtain labels for unlabeled training examples. Aimed at domains in which a large quantity of data is available but labeling is expensive, active learning directs the expert to label the more informative examples with the goal of minimizing the number of queries. This is directly relevant to the LfD problem—since getting demonstrations from a human teacher is inherently expensive, we want to maximize the utility of demonstrations making the most efficient use of the human teacher's limited time. In this section we cover approaches for applying Active Learning in an LfD setting.

As motivating evidence, the work of Chao et al. compared passive supervised learning to active task learning and addresses the question of *when* to ask questions in a mixed-initiative AL setting [45, 60]. These works found that active learning outperformed passive supervised learning, and users preferred AL (e.g., they thought the robot was more intelligent when it asked questions). However, their experiments highlighted the need for understanding how to balance a mixed initiative interaction, particularly how to determine when the robot should ask questions versus let the human provide any examples they like. Rosenthal et al. investigate how augmenting questions with different types of additional information improves the accuracy of human teachers' answers [222]. In later work, they explore the use of humans as information providers in a real-world navigation scenario [223]. Thus we have evidence that designing robot learners that ask questions is a good idea, but all questions are not equal since some will be more informative and useful for the robot.

Inspired by these factors, Cakmak introduced *embodied queries*, a spectrum of three different query types for robot active learners [44, 48]: label queries, demonstration queries, and feature queries. We give a summary of each of these in the next three sections.

6.4.1   LABEL QUERIES

The conventional *query* in the AL literature involves choosing an unlabeled instance and requesting a label for it, e.g. "What action should I perform here?" The instance can be chosen from a pool of unlabeled instances or instantiated by the learner in some way. Such queries have been used in learning skills on a robot, where a skill is represented as a policy that maps a state to a discrete action. Chernova and Veloso introduce a classification-based technique that uses measures of similarity to past examples and classification confidence to determine when the robot should act autonomously or stop to request a demonstration [66]. The demonstration is provided in the modality of the original demonstrations for an action, in this case through a GUI. Lopes et al. [167] introduce an active learning technique for inverse reinforcement learning, in which the learner is able to query the demonstrator about the reward value of specific states.

Many robot skills involve continuous actions and the input from the teacher to the learner is a sequence of state-action pairs (i.e. trajectories). In these cases it is impractical to ask for an isolated state-action pair (e.g., asking for the motor commands of a given arm configuration), Cakmak addresses this by having the robot execute an entire motion and ask whether the skill was performed correctly. Thus, a label query on the entire demonstration, essentially asking the teacher "Should I include this trajectory in my set of positive examples of this task?" Methods for generating label queries depend on the particular framework used for representing the skill. However, a general approach applicable to most frameworks is to sample trajectories from the learned skill and evaluate them with a certain criterion. For instance, the robot can choose to query the trajectory that it is least certain about or the trajectory that is most likely to increase the applicability of the skill.

The information provided by this kind of query depends on the answer. If the response is positive, then the motion can be used as another demonstration. What to do with negative examples is not as straightforward. LfD methods are designed to encode a skill from positive examples only. The space of "what not to do" is so much larger that it is not typical to try to model the negative class explicitly. One way to make use of negative examples that arise from label queries, is to update the learned model such that the probability of the negative data being generated by the model is minimized while the probability of the positive data being generated is maximized. The main issue with this idea is the attribution of negativity to the whole trajectory, while only parts of the trajectory might be responsible for the failure. A second approach for making use of negative examples is to use them to guide the learner's future queries towards positive examples. As an example of this, Grollman and Billard explicitly make use of failed demonstrations by assuming that they are likely to be a "near miss" [101]. By building a model of the failed demonstrations their system can generate new exploratory trials in the neighborhood of this model but explicitly *not* what the human showed. Thus the human's failed demonstrations are a seed for the robot's exploration to build a successful model.

6.4.2   DEMONSTRATION QUERIES

The second type of query in Cakmak's framework is a demonstration query, where the robot finds a configuration of the environment that its model does not cover, and asks for a demonstration from here. These are analogous to a method known as *active class selection* [166], which consist of requesting an example from a certain class.

In demonstration queries, the learner only specifies certain constraints, while the trajectory is still produced by the teacher. As a result, the learner has less control over what information is acquired than in label queries. One way to constrain trajectories provided by the teacher is to specify the starting state. Since trajectories are often represented with a sequence of end effector configurations relative to a goal object frame, the robot can configure its end effector in a certain way relative to the goal and request the demonstration. For example, in [100], the robot actively selects points outside the region of stability of a learned policy, and requests demonstrations from these states. A different way of constraining the demonstrations provided by the teacher is to allow the teacher to control only a subset of the robot's joints while the robot executes a certain trajectory on the rest of the joints, as in [53].

6.4.3   FEATURE QUERIES

The final query type in Cakmak's taxonomy is a feature query. In this query type, instead of asking what to do, the robot asks about the variance or invariance of particular features of the action ("is the location of X important in this situation?"). This is inspired from a technique in AL that involves asking whether a feature is important or relevant for the target concept that is being learned [87, 212], which has been successfully applied in document or email classification. Whereas label queries would typically involve the teacher reading documents one by one and providing category labels for them, which is cumbersome even in an AL setting, a feature query directly asks whether a word is a strong indicator of a certain category. This allows the learner to directly modify its model for categorizing new documents and drastically reduces the time spent by the teacher.

The crucial element in the success of feature queries in these document text examples is that the features (words) of each instance (document) are meaningful for humans, and the way the features contribute to the classification problem is intuitive. Robot skill learning is in a unique

Figure 6.4: Examples of label, demo, and feature queries [48].

position to take advantage of this method: while features might not be as human legible (i.e., feature names might be too technical and feature values might be arbitrary numbers) the robot's embodiment can be used to *show* the features instead of referring to them by name.

Methods for choosing feature queries are also dependent on the framework used for representing and learning skills. One framework that allows feature queries to be directly integrated is task space selection [124, 179]. This involves representing demonstrations in high dimensional spaces that involve features that might or might not be relevant (e.g., relative position and orientation of different points on the robot to different objects in the environment). Methods try to identify a subspace or assign weights to each feature such that the skill is best represented. In this context a feature query is to directly ask whether a feature is important for a skill. These queries can also be used for directly manipulating the skill representation or for guiding other queries.

Cakmak showed this in the context of a humanoid robot learning object manipulation skills, where the underlying representation was analogous to a Gaussian Mixture Model [44]. In this case, executing a feature query for the learned task involves generating a new skill trajectory that perturbs a single dimension of one of the underlying Gaussian models in the skill. An interesting aspect of this work was the use of the robot's embodiment to show features rather than ask about them. For example, in a pouring skill, the robot steps through the skill and stop at the point at which it want to make a query, and then selecting the x-rotation as the dimension to query it shows a range of values around the median value that its model would suggest while asking the user "At this point in the skill, does *this* matter?" The alternative would be a question of the form:

"At this point in the skill, does rotation about the x-axis matter?" Thus, making use of the robot's embodiment during the query can greatly improve the interaction.

6.5   SUMMARY

Our discussion of tightly coupled interactions within LfD brings us full circle by highlighting the powerful role that the human teacher can play in robot learning. Very few of the algorithms and approaches introduced in Chapters 4 and 5 have been utilized in an iterative or incremental fashion, and even fewer with a social interaction with end-users in mind. Combining the ideas presented in these three technical chapters offers some of the most promising areas of future work for the design of LfD systems. Having covered the technical chapters of this book, we now go on to present evaluation methods for interactive LfD studies.

CHAPTER 7

Designing and Evaluating an LfD Study

So far in this book, Chapters 2–6 have walked through the entire LfD pipeline, surveying a variety of state of the art algorithms, systems, and approaches. In this chapter, we turn the discussion to the topic of evaluating such techniques. The chapter covers, very briefly, the general protocol for conducting experiments with human subjects for the purpose of evaluating LfD systems. We step through an example of designing, running, and analyzing an LfD HRI data collection experiment. This is meant as a very introductory version of experimental design, assuming no background in HCI or human factors, and focusing on the simplest type of study that one might run for LfD evaluation. Our goal is to present a tutorial aimed at AI or Robotics researchers that want to begin validating their systems with end-users. At the end of the chapter, we refer the interested reader to additional resources that can provide more depth on the interesting complexities of conducting HRI research. Our discussion will center on the following hypothetical scenario:

> Researchers Sally and Bob have just completed the design and implementation of their Amazing Task Learning (ATL) algorithm, which takes demonstration input as teleoperated or kinesthetically provided experiences of the task. As the designers of this algorithm, they have become expert demonstrators, and are able to achieve great success with the ATL algorithm on the humanoid robot in their lab. Characterizing the learning performance of the system with expert demonstrations is one type of evaluation, and can be a nice way to systematically demonstrate various components of the system or algorithm. In addition to this, they have decided to perform an HRI study to see how ATL performs with people other than the designers in order to show the generality of the approach.

7.1 EXPERIMENTAL DESIGN

One of the main goals of an Human-Robot Interaction (HRI) study is to verify and validate the performance of a designed robot system. The primary goal of fielding an interactive robot system with people other than the system designers, is to be able to make claims about the generalization of the designed interaction to the broader population. This is precisely the motivation for testing LfD systems with an HRI study.

While we will focus in this chapter on the specific LfD evaluation scenario described at the beginning of this chapter, what we cover here is generally applicable to any experimental design. The overall goal of our LfD experiment, and any experiment, is to understand and predict the relationship between variables. And the goal of good experimental design is to isolate the variables of interest such that we can draw concrete conclusions about their relationship. As such, an experiment has two classes of variables.

- **Independent Variables**: The goal of a study is to test certain hypotheses about the relationship between independent and dependent variables. The independent variables are aspects of the experiment that the experimenter will purposely manipulate in order to test these hypotheses. In the context of LfD, one example would be testing two different versions of the ATL algorithm, or to test ATL versus some other baseline algorithm. In this case, algorithm version is an independent variable the experimenter is manipulating.

- **Dependent Variables**: These are aspects of the experiment that can be measured in order to see the effects of the manipulation of the independent variable(s). The nature of the dependent variables is related to what claims the researcher wants to make about the effects of the independent variable. In general for LfD systems there are two classes of dependent variables to consider. You will want to use standard Machine Learning metrics (e.g., convergence rate, sample complexity, task performance) in order to make claims about how your appropriately designed learning interactions result in learning performance gains. In addition to this, you may also want to use observation and survey methods to measure aspects of the interaction itself, the human teacher's behavior, and the teacher's subjective perception of the teaching/learning experience. More details about metrics are covered in the following sections.

The first step in designing a study is to formulate the hypotheses in order to decide what independent and dependent variables are necessary to address them. The hypotheses will be of the form: "I expect that changing [insert independent variable] will have an impact on [insert dependent variable]" For our ATL experiment, we are interested in having humans teach a robot as it runs different interaction modes or different versions of a learning algorithm. This allows us to draw conclusions about how the implemented changes impact the teaching/learning process. For example:

H1: We expect that the ATL algorithm will have better learning performance than the Baseline, in terms of learning rate and sample complexity.

H2: We expect that teachers will find kinesthetic teaching to be more satisfactory than teleoperation.

The type of the experiment can be described by the number and type of independent variables in the design, illustrated in Figure 7.1. *Two group design*: the simplest design is to have a

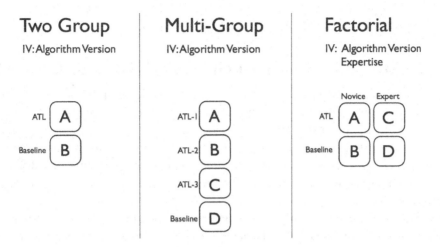

Figure 7.1: The number and type of independent variables (IV) defines the experiment type.

single independent variable, testing two different conditions of this variable. For example, comparing the ATL algorithm (group A) to the Baseline algorithm (group B) yields a two-group design. *Multi-group design*: the next level of complexity is to test several conditions of a single independent variable. For example, testing four different parameter settings for the ATL algorithm would be a multi-group design, one group (A, B, C, and D) for each manipulation of the independent variable. *Factorial design*: if there are multiple independent variables that are manipulated at once this is called a factorial design. This is necessary when the hypothesis is that there is some interplay between the two manipulations. As an example, we may think that the expertise level of the teacher (novice, expert) will impact their ability to use the ATL algorithm versus the Baseline. So we have two levels of the expertise variable that we would like to test and two levels of the ATL variable to test. This gives us a 2x2 design, yielding four experimental groups each with a particular setting of the two independent variables (e.g., A: novice-ATL, B: novice-Baseline, C: expert-ATL, D: expert-Baseline).

Given the design of the experimental conditions (or groups), the next question is how participants will be assigned to those groups. One option is a *between-subjects* design. In this case each participant in the experiment is assigned to only one of the experimental conditions (i.e, each participant only does one of A, B, C, or D for any of the designs shown in Figure 7.1). The other option is a *within-subjects* design, where each participant does all the experimental conditions (e.g., in the two-group design, each participant does both A and B). A factorial experimental design can also have a *mixed* design, where one independent variable is assigned between-subjects and another independent variable is assigned within-subjects. The choice of between vs. within subjects design is going to depend on the particular experiment, domain, task, and measures. But the decision usually comes down to which choice will result in a better controlled experiment, more on this later when we discuss experimental controls (Section 7.4).

7.2   EVALUATING THE ALGORITHMIC PERFORMANCE

Having determined the groups in the experiment, the next design question is how best to measure the differences hypothesized between these groups. In this section, we discuss the use of standard Machine Learning performance measures as dependent variables expected to reflect differences across the experimental groups. In addition to this, one may want to claim something about the interaction itself, to show a difference in teacher behavior across groups, or a difference in their subjective account of the interaction. Techniques for measuring these changes are discussed in the following section.

Differences in performance between different algorithms can be measured in a number of ways. One of the most important considerations is the accuracy or success of the resulting behavior at the task being learned. *Accuracy* is reported as the proportion of actions that are correct; its inverse, the *error rate*, reports the number of incorrect actions. Unfortunately, it is often difficult or impossible to measure the accuracy rate in robotic domains because the correct action is not known for an arbitrary state. Consider, for example, teaching an autonomous helicopter to perform aerial acrobatics. It is much easier to determine whether a particular maneuver has succeeded than whether the correct control input was provided at every timestep of the trial. In such cases, the *success rate* can be used to report the proportion of successful completions of the target task.

In addition to measuring how well the learned model performs, it is important to compare what it takes to achieve that level of performance. The primary measure of importance is the number of training samples required for training. From a theoretical perspective, *sample complexity* provides a high probability upper bound on the number of timesteps required by the algorithm to learn the policy for tasks in some class. For algorithms that refine the policy over multiple iterations, such as reinforcement learning, the *convergence rate* of an algorithm, averaged over multiple runs, is often reported as an empirical estimate of sample complexity for a particular domain.

Additionally, within the context of LfD, it is important to characterize the teacher's input. At the very least, results must report the number and type of interactions the teacher had with the robot (e.g., number of demonstrations, number of corrections). If possible, the quality of the demonstrations should also be measured. This can sometimes be done by measuing how well the human teacher can perform the target task, how long the teacher takes to perform it, or how consistent the demonstrations are from run to run. These values provide insight into the quantity and quality of teacher input required to train the algorithm in question, and the resulting algorithmic performance. In the following section we now discuss techniques for evaluating the interaction from the teacher's perspective.

7.3 EVALUATING THE INTERACTION

The options for measuring the interaction itself fall into two categories: subjective and objective. Subjective measures are usually the first thing that comes to mind when one thinks of doing experiments with human subjects. A common misconception with running an HRI study is that it is sufficient to have a person complete some interaction with a robot and then give them a survey to measure their subjective experience, hopefully seeing some difference in their answers across the conditions. But it is important to not overlook the importance of supporting that claim with objective measures.

As an example, the subjective measure of asking people "how frustrating" they found the robot interaction can provide data to support a claim like: "our ATL Algorithm was significantly less frustrating for the human teacher, compared to the Baseline Algorithm". Now to support and provide insight into this conclusion it is useful to also measure objective aspects of the interaction that could point to *why* there was a difference in some subjective measure. Continuing the same example, "We see that participants in the ATL condition had significantly fewer errors and completed the teaching task faster." While the exact choice of metrics will depend on the particular experiment and hypothesis, it is a good rule of thumb to include both subjective and objective measures to support your claims. A 2006 survey of common metrics for HRI [240] provides a useful overview of standard practices in the field.

7.3.1 SUBJECTIVE MEASURES

Subjective measures include any measure in which you ask the participant to report their subjective experience of the interaction. A common way to collect such data is in the form of a survey, where participants answer questions about the interaction. Answers are usually in the form of a rating scale (also called a Likert-scale [160]), or a forced choice between options. The most common form of a Likert questionnaire item is a 5-point rating scale, where respondents indicate their level of agreement or disagreement on a symmetric scale for a series of statements. For example, the wording of a typical five-level Likert item could be:

```
1 - Strongly disagree
2 - Disagree
3 - Neither agree nor disagree
4 - Agree
5 - Strongly agree
```

Subjective measures in general are prone to various types of distortion. Some respondents may avoid using the extreme response categories (known as a central tendency bias). Another type of bias is the social desirability bias, where respondents may tend to try and portray themselves or their organization in a favorable light. As an example in our HRI LfD setting, people may be reluctant to make very negative comments about either ATL or the Baseline algorithm, making it hard to draw strong conclusions about the comparison.

Because of these issues with bias in subjective data, the precise wording of a survey question is very important. Designing a good questionnaire that captures the desired dependent measures is a research question in and of itself. Thus, in choosing subjective measures, whenever possible one should try to reuse a questionnaire from related work that has been shown to be a reliable measurement tool. Some examples include the following.

- Anxiety: The Negative Attitudes toward Robots Scale (NARS) has been put forward as a tool for measuring people's anxiety toward robots [195]. The test has been used several times in HRI studies since its introduction in 2003 (e.g.,[22, 258]), and requires participants to respond to statements such as "I would feel uneasy if I was given a job where I had to use robots."

- Task workload: NASA-TLX is a subjective workload assessment tool that allows developers to perform subjective workload assessments on operator(s) working with various human-machine systems [108]. NASA-TLX is a multi-dimensional rating procedure that derives an overall workload score based on a weighted average of ratings on six subscales, which include Mental Demands, Physical Demands, Temporal Demands, Own Performance, Effort, and Frustration. The test has been used in a wide range of robotic studies [86, 93, 238], and includes questions such as "How hard did you have to work to accomplish your level of performance?"

- Trust: The Working Alliance Inventory [112] is a questionnaire commonly used in therapy and other helping relationships that tracks trust and belief in a common goal that the therapist and patient have for one another. The form solicits responses to statements such as "We are working towards mutually agreed upon goals." The questionnaire has been used to evaluate trust between the user and the robot in a variety of robotic and software agent applications [28, 132]. A broader survey of techniques developed by varying disciplines for measuring trust can be found in [266].

7.3.2   OBJECTIVE MEASURES

In an HRI setting we have the benefit that the robot is in a perfect position to log any number of metrics about the interaction that can serve as dependent variables in the experiment. For example: number of demonstrations provided, length of demonstrations, number of commands given, number of queries generated by the robot, etc. Objective measures about the learning performance are likely metrics to use, such as sample complexity and precision/accuracy of the learned model over time if learning is occurring incrementally during the interaction. The great benefit of these types of dependent variables is that they are easy to collect and non-intrusive, since it is just part of the interaction itself. These reflect the actual experience people had during the interaction, not their self-reflection or interpretation of what happened, thus their behavior speaks for itself. For example, on a survey people may be positive toward both ATL and the Baseline algorithm, but we may be able to show that ATL is better when you look at learning performance, sample complexity, number of errors, etc.

If some behavioral measure of interest cannot be logged automatically during the interaction, then another option is to record the interaction on video and have a third party observer log the behavioral metric of interest by annotating the video (also called *video coding*). To do this we first define the measured variable of interest and exactly how a person annotates it, this is the *coding protocol*. For example, a person (*coder*) watches video of human and robot interacting and marks the every time there is eye contact, or every time that the human talks to the robot, or every time the human laughs. This is a common type of dependent measure in psychology and linguistics research, hence various tools have been developed for aiding in the process of annotating video/audio, one example is ELAN.[1] Ideally, when using this type annotation as a dependent variable, multiple coders would annotate the data. In this case it is important to confirm the reliability of the coding process with the Cohen's kappa coefficient, which measures the amount of agreement between the coders in their annotation. In practice, having multiple people code the entire dataset is quite costly. An alternative is to have a single coder for the entire dataset, but confirm reliability by having multiple additional people code different small portions of the data to assess agreement (via the Cohen's kappa metric) and thus reliability. It is also typical to have multiple people code different portions of the dataset that overlap slightly, and then the reliability measure is calculated on these overlapping portions to confirm the reliability of the annotations. For a detailed example of annotating behavior in video and performing statistical analysis see [78].

7.4   EXPERIMENTAL CONTROLS

The researcher's goal in designing a study is to be able to make conclusive statements about the relationship between the studied variables. In particular, the experimental design sets up a very specific claim to be made about how the independent variable(s) effect the dependent variables. This kind of claim can only be made if a convincing argument can be made that the only thing that distinguishes the different groups in the experiment is the manipulation of the independent variable. Ideally, all other factors in an experiment will be controlled (accounted for by the control measurements) and none will be uncontrolled, and the extent to which this is true strengthens the claim that the independent variable is what causes some difference in a dependent measure.

The first set of controls to think about are the experiment task itself. Participants in each group should be asked to do exactly the same thing in the same environment, with the only difference being the manipulation of the independent variable. For example, in our LfD experiment the set of tasks that people teach the robot should be the same across experimental groups. The experiment should take place in the same environment, with the same workspace setup.

One factor of the experimental task that is important to control, is the experimenter herself. The experimenter will probably be required to give instructions to the participant about what to

---

[1]`http://tla.mpi.nl/tools/tla-tools/elan/` "ELAN is a professional tool for the creation of complex annotations on video and audio resources"

do, or perhaps intervene at particular times in the experiment for one reason or another. It is a good idea to *write down* a protocol for the experimenter to follow in each of these expected interactions with the participant. This will help to ensure that every participant received the same amount and type of instruction on the task, and that participants in one group were not biased in some way by the experimenter.

Novelty effects can be particularly important for LfD and HRI. In many cases participants may have never seen or interacted with a robot before. We don't want the effects of this first interaction (so-called novelty effects) to completely overpower the manipulation of the independent variable in our study. One common way to try and control for novelty effects or reduce their impact on the experiment is to have a practice interaction with the robot. For example, in our ATL experiment, we may have people teach the robot a simple practice task in order to familiarize themselves with the robot, the workspace, and the general experimental protocol. The data from this practice task would not be recorded or used in our analysis.

Controlling for practice and familiarity is essential in a within-subjects experimental design. Remember that this means that a single participant will perform multiple experimental conditions. In our two-group example, a within-subjects design means that each participant teaches the robot with the ATL algorithm and also the Baseline algorithm. In this case our data would be biased if the ATL or Baseline were always the first algorithm they interact with, since we would expect their performance as a teacher to get better with each interaction/demonstration. This is addressed by "counter balancing" the conditions in a within-subjects design, ensuring that each ordering of conditions in the experiment is seen uniformly in the data. In our case, we would have half of our participants teach with ATL first and Baseline second, and the other half would start with Baseline first. If the effect of doing one condition before another will be too significant, this is often a reason for deciding to use a between-subjects design. Thus, the between-subjects design is controlling for the effect of familiarity with the experiment or the task.

Typically, the decision of between- or within-subjects is not clear cut, there is a choice to make. Within-subjects design is preferred when possible, since it requires fewer participants to collect the same amount of data. It also has the advantage that the variance due to individual differences or natural dispositions is controlled across the groups, since each person (with their individual differences) is contributing data to all of the experimental conditions. However, it is not always possible to use a within-subjects design. One reason might be the familiarity effects mentioned above. Another could be simply that asking each subject to complete all of the experimental conditions may be too demanding. In this case, the effect of task fatigue that a participant may have by the end of the study might impact the quality of the data. In a between-subjects design, variations that cannot be controlled should be evenly distributed across the experimental groups (e.g., controlling the distribution of factors like age, gender, etc.). With a large number of participants, randomly assigning participants to experimental groups is the typical way to guarantee this distribution. With smaller numbers, then explicitly balancing the groups by gender, or age group is appropriate.

7.5   EXPERIMENTAL PROTOCOL

Now that we have a nicely designed experiment, it's time to run it! The first step is to apply for ethical approval with our institution. For example, in the U.S. this is Institutional Review Board (IRB) approval, and in the E.U. these are called Ethics Committees. This is required for any experiments that involve human subjects. Apprehension about this institutional approval process is probably the number one reason that we do not see more experiments with naïve subjects in the LfD research community. So our goal is to dispel the myth that getting approval is hard. In the appendix, we have included an example IRB protocol and consent form for a generic LfD study. Of course the exact process depends on the particular institute, but given our collective experience with IRB at three different U.S. academic institutions, we believe that a typical LfD study is quite straightforward from an ethics approval perspective. These experiments often fall under the category of "minimal risk," allowing the application to go through an expedited review process instead of a detailed one.

Once approval is granted by the ethics committee, it is time to collect data. How to solicit participation in the study will depend on the research community. In a university setting, it is typical to advertise via posters, email and word of mouth to recruit students as participants. In a corporate setting, there can be resources for recruiting focus groups from the local community. However, people are recruited to participate in the experiment, it is important to keep in mind the generality goal that we started with. If a study is conducted exclusively with robotics PhD students, then this is the extent to which we can claim our approach generalizes. The fields of both engineering and social science tend to suffer from this kind of convenience sampling (i.e., studying participants that are easy to recruit). Whereas if we explicitly recruit participants with little to no experience programming robots or with little to no Machine Learning background, then the claims about how our LfD system generalizes can be stronger. In general, if we recruit a diverse group of participants from the target population, the claims of our HRI study can be more broad.

The following are the detailed steps of running our experiment for a single participant.

1. Setup: Before the participant arrives, we set up the environment, workspace, and robot, paying close attention to our controls and the experimental condition for this participant.

2. Introduction: We have written down the instructions we will read aloud to participants, to help make sure to say the exact same thing to everyone. One could also decide to write down instructions and just have people read them, but in our experience this tends not to be as effective as talking through the instructions. After the explanation, we ask if they have any questions and make sure that they are clear about the instructions before moving on. If someone goes into the experiment with a misunderstanding, this would negatively impact the validity of our experiment. So it is important to take the time to make sure they understand.

3. Informed consent: After we have given them all the necessary information to make the *informed* decision to participate, we have the subject sign the consent form.

4. Collect pre-experiment measures: Any dependent measures that require pre- and post-experiment measurement should be collected immediately prior to the interaction. For example, with the subjective measures collected by questionnaire the hypotheses may require administering the survey before and after the interaction, in order to compare differences.

5. Robot interaction: Next, the participant interacts with the robot according to our designed experiment. Unless it is part of the study plan, we try to intervene as little as possible during this data collection phase, so as not to introduce any bias. However, even if the experimenter is diligent about not initiating any interaction with the participant, often the participant may turn to the experimenter to ask what they should do at some point. It is good to expect this and plan for it. We try to either give everyone the same specific advice to particular questions that come up often, or a generic response like "just do what you think is best, pretend I'm not here and you're teaching this robot on your own."

6. Collect post-experiment measures: Immediately following the interaction, we administer any post-experiment subjective measures.

7. Exit interview: After having run the planned experiment and collected all of the data, it is a good idea to have an informal exit interview to debrief the participant. At the very least this is the time to explain a little more about the hypothesis, and explain which experimental condition they were in (if they only saw one). Additionally, this is a great opportunity to get informal feedback about what they thought of the interaction with the robot, or how they approached the task of demonstrating. Sometimes this informal feedback can have a huge influence on directions for future work, even if it is not explicitly part of the results reported for this experiment.

7.6   DATA ANALYSIS

The final topic to address is techniques for drawing conclusions about the hypotheses given the data now collected. The goal is to determine whether or not the dependent variables change because of the independent variable(s).

7.6.1   CHOOSING THE RIGHT STATISTICAL TOOL

There are two general categories of techniques to use to look at the data: descriptive and inferential techniques. The goal of *descriptive techniques* is essentially to provide a summary of the data. It can be informative to look at the means and standard deviations of each continuous valued dependent variable. Note that the mean is not what we want to consider for questionnaire data with a Likert scale. In this case, the data are ordinal with an inherent ordering, but the difference between

Table 7.1: Summary of the appropriate statistical inferential tools to use depending on experimental design and the nature of the dependent measure

	Parametric methods	Non-parametric methods
2 groups	Student's t-test	Mann-Whitney test (between-subjects)
		Wilcoxon signed rank test (within-subjects)
3+ groups	ANOVA test	Kruskal-Wallis test

strongly disagree and disagree is not the same as the difference between neutral and agree, thus an average is not meaningful in this context. A more meaningful descriptive statistic for this type of data is the median value or the mode. It is a good practice to begin data analysis by looking at descriptive statistics of the data before using any inferential techniques.

The goal of *inferential techniques* is to investigate the differences between the conditions of the experiment. After plotting the means and standard deviations of the dependent measures across the groups, hopefully we are able to *see* that the data seems to change depending on the group. Inferential statistics are about determining whether or not this difference is significant such that a concrete conclusion can be drawn.

While the particular statistical test to perform depends on the nature of the data and the groups, the test for statistical significance is always set up in the same way. To show that there is a significant difference between our experimental groups we prove that it is unlikely that the data were all drawn from the *same* distribution. This allows us to conclude that in fact the groups represent *different* distributions. Thus, the null hypothesis for an inferential test is: "all of my data were drawn from a single source," and if we can show that this is statistically unlikely, then we can conclude that the data does indicate that our experimental groups are different. It is a proof by contradiction. In the remainder of this section we will cover how to decide which method is appropriate for particular data. We will not cover the details of particular methods here, but instead refer to additional resources at the end of this chapter.

For dependent measures for which the underlying data is real-valued, we make use of parametric methods. These methods look to model the distribution of the data (e.g., as a Gaussian with parameters μ and σ) and infer whether or not there are significant differences between the distributions of data from the different conditions. The method to use depends on the number of groups (see Table 7.1). It also depends on whether or not the samples were taken within group or between group. When participants in the experiment contributed data for more than one experimental group, then data is said to have *paired samples*, i.e., there is a sample in group 1 and a sample in group 2 that are related in that they were both collected from the same subject. In this case, the paired samples version of the Student's t-test and ANOVA test are appropriate.

When the underlying distribution of data for a dependent measure is not Gaussian (e.g., a measure that is ordinal/nominal, or when a discrete or continuous data violates assumptions of normality), then a nonparametric statistical method should be used. A common type of dependent measure that requires this sort of analysis is the data from a Likert survey. As discussed previously, the mean and standard deviation are not very meaningful for ordinal data. Instead we want to investigate the differences between the medians of these data across the groups. The method is different depending on the number of groups, and whether or not there are paired samples (see Table 7.1).

7.6.2   DRAWING CONCLUSIONS

The result of the inferential method will be a statistical value (e.g., the t in a t-test, or F in an ANOVA) and the p-value. The value p is the significance level. It tells us whether or not we can infer that there is a *significant difference* between our groups with respect to the dependent measure in question. If so, it is valid to reject the null hypothesis, concluding that it is false. Actually a low p-value means either that the null hypothesis is false, or that the null hypothesis is true and a highly improbable event has occurred. The rule of thumb is that with a value of $p < .05$, it is appropriate to conclude that the null hypothesis is false, that there is a significant difference between your experimental groups. In system-evaluation studies, like the LfD HRI studies we are describing here, "marginal effects" are also usually interesting and commonly reported as *trends* for slightly larger p-values ($p < .1$).

This insight into what the test says about the null hypothesis is important, because it helps us avoid a common mistake. It can be tempting to conclude that the null hypothesis is *true* when $p > .05$. But this only makes sense in a world where there are only two alternatives, i.e, A : my change to the independent variable caused a change in the dependent variable, B : my change to the independent variable did not cause a change in the dependent variable. In reality, there are several alternative hypotheses, therefore $\neg A$ is not equivalent to B. Hence, if a test results in $p > .05$ then the test is inconclusive about the null hypothesis. It does not allow one to conclude there is a significant difference, but also does not allow a conclusion that the groups are similar. A regression analysis, showing how factors across the groups are correlated, is one way to draw conclusions about the similarities between experimental groups, details of regression analysis are beyond the scope of this chapter but can be found in [16].

If a statistical test finds a significant difference across the experimental groups for a dependent measure, this is called an "effect" of the independent variable. For example, in our two-group study of the ATL vs. Baseline algorithm, say we find a significant difference between the groups in the accuracy measure of the algorithm performance. Then we can conclude that "the algorithm version had an effect on learning performance, ATL was found to be significantly more accurate than the Baseline."

In a factorial design, with multiple independent variables, there are two different kinds of effects to look for: main effects and interaction effects. Each of the independent variables has its own main effect on the deponent measure in question, and then there can also be an interaction effect between each of the independent variables. These effects are best seen with an example.

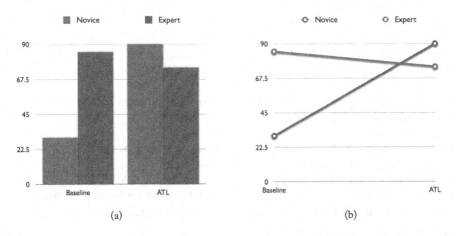

Figure 7.2: Two different visualizations of example data from the four groups of the factorial design in Figure 7.1: novice-ATL, novice-Baseline, expert-ATL, expert-Baseline. Plotted is the average for each group for the dependent measure "accuracy of the learned model." This data shows a main effect for each independent variable as well as a crossover interaction effect between the two.

Consider our factorial design from Figure 7.1 with the two independent variables of algorithm version (ATL, Baseline) and expertise (novice, expert). Assume that the recorded data, seen in Figure 7.2, shows two different visualizations of the data from the four groups for the dependent measure of the accuracy of the learned model. Looking at the bar graph in Figure 7.2(a), the main effect of algorithm is shown by the fact that the two bars under ATL are higher on average than the two bars for Baseline. Similarly, it looks like there is also a main effect for expertise, since the average of the two expert bars is higher than the average of the novice. Interactions are effects between the independent variables of a factorial experiment. If one independent variable depends on the other there is an interaction effect, if one variable is the same regardless of the other then there is no interaction. A particularly interesting interaction is the crossover interaction, where the effect of one independent variable is reversed depending on the condition of the other independent variable. As seen in Figure 7.2(b) this is what we have in our example data. The ATL algorithm improves accuracy of the learned model over the Baseline for novice users, but not for experts. These three different effects for a 2x2 factorial design are three separate analyses and claims to be made. A particular 2x2 experiment can result in any combination of these effects being significant or not.

7.7   ADDITIONAL RESOURCES

In this chapter, we have provided an overview of the most common methodologies for experimental design and evaluation as related to LfD studies. This area is supported by a large body of literature that provides further details on related topics. Here we list several resources that are particularly relevant to LfD study design.

- "An Introduction to Human Factors Engineering" by Wickens et al. [269] provides a broader coverage of the experimental design considerations introduced in this chapter.

- Hornbaek [111] provides a similar introductory tutorial in the context of Human-Computer Interaction experiments.

- For an in-depth explanation and derivation of the statistical analysis tools and techniques suggested in this chapter, [16] is a good introductory text.

- Steinfeld et al. [239] survey common task-based metrics (dependent measures) seen in various HRI settings and tasks, from navigation to manipulation to social interaction.

- Olsen and Goodrich [197] provide a similar survey for the specific context of shared autonomy, which is particularly relevant to a Learning from Demonstration scenario where teleoperation and shared autonomy are a popular input device.

- Bethel et al. [27] provides a concise introduction to the use of physiological measures in HRI studies with the goal of measuring a person's response to the robot with which they are working.

CHAPTER 8

Future Challenges and Opportunities

With the aim of being forward looking, this book both reviews existing work in Learning from Demonstration (LfD) and highlights future challenges. Each of the chapters of this book has offered some suggestions for future research in the respective topics; in this chapter we discuss three closely inter-related research directions that we consider major challenges for the field. We also provide the interested reader with a list of additional reading resources related to LfD.

8.1 REAL USERS, REAL TASKS

The first challenge area for the field of Learning from Demonstration is to incorporate more development and evaluation with real users performing real-world tasks. Just as in many other areas of robotics, LfD experiments are typically performed in constrained lab environments, often with the algorithm's developers standing in for the target population of end users. While this type of testing is an important preliminary step, evaluation must also be carried out with external users, preferably in environments that resemble the target use case.

Research has shown that LfD algorithms evaluated only with input from their developers, without iterative testing with users, can fail to perform as expected in user trials [242]. Instead, policies that might take under a minute to learn with a graduate student programmer for a teacher become impossibly complicated due to the numerous errors made by a novice teacher, due to misconceptions about the algorithm. Understanding the behavior of teachers and the types of inputs they provide is critical to successful deployment of LfD techniques. This is exemplified in a series of papers by Thomaz and Breazeal, in which the authors first conducted a study to understand human behavior during teaching and then developed a learning algorithm that would leverage the teachers' natural behavior in the most effective way [252, 253]. More algorithms aimed at *real users* is a challenge for the field.

Along these same lines, is the challenge of *real tasks*. A highlight in this area is the joint work of Ratliff et al. [217] and Silver et al. [232], which demonstrated the use of LfD in controlling an unmanned ground vehicle for navigation across highly complex off-road terrain. One of the greatest barriers to working with complex domains and real users are the associated time and effort demands on the developer. Emerging interactive technologies, such as remote robot laboratories [199] and web-based frameworks for conducting HRI studies over the web [255], have the potential to lower these barriers by lowering the cost of conducting complex user studies in the future.

8.2 HRI CONSIDERATIONS

Closely related to the above challenge is the issue of greater consideration of human-robot interaction factors within Learning from Demonstration. The vast majority of the research in LfD has emerged from traditional Machine Learning approaches, focusing primarily on algorithmic performance evaluated based on learning time, number of demonstrations and task accuracy. While algorithmic performance is critical for the success of LfD, we argue that greater consideration must be given to the usability of the developed techniques as the field matures. We hope that HRI evaluation methods will see increased adoption within this research community in the coming years, in addition to continued focus on learning performance measures and task outcomes.

8.3 ADVANCING LEARNING THROUGH BENCHMARKING AND INTEGRATION

Closely tied to the above evaluation methods is the issue of benchmarking and algorithm comparison. Unlike many other areas of robotics and machine learning, Learning from Demonstration as a field lacks standard data sets or domains that can be used to systematically evaluate the developed techniques. One of the reasons for this is the breadth of problems being considered under the LfD umbrella, and the wide range of differing assumptions that algorithms operate under. However, even if a common set of benchmark tasks could be defined, the more significant obstacle is the reliance on diverse forms of human input, the need for which has precluded the development of a standard testing procedure or domain. As a result, the vast majority of existing techniques have been applied to only a single, often unique, domain. The open challenge that remains for the field is how to effectively advance the state of the art to tackle more complex, real world tasks. This includes challenges such as scaling to larger state spaces, integrating multiple skills learned from LfD into more complex behaviors and combining the individual pieces currently being studied within the field, such as feature selection, active learning, goal learning and various forms of user input. This topic continues to spark debate at symposia and workshop gatherings, but no established solution currently exists of which we are aware.

8.4 OPPORTUNITIES

It is a very exciting time to be working in the field of Learning from Demonstration. There is currently a surge of interest in adaptive robots in real human environments, like the manufacturing setting (e.g., with robotics like Baxter from Rethink Robotics). Learning from end-users is envisioned as the key to success in these domains. Similarly, the DARPA Robotics Challenge exhibits the feasibility of really getting humanoid robots to operate in dynamic human environments someday soon, where there is an opportunity for LfD to have a great impact on the utility of these robots. And more generally, there is a recent run toward low-cost mobile manipulation platforms that could very soon make working with real robots not be something that is reserved for well-funded labs at research institutes and universities. Getting more robots in the hands of more researchers will be an opportunity for many more people to get involved in the field of LfD than ever before.

8.5   ADDITIONAL RESOURCES

There are a number of additional resources that readers may find useful and complementary to the content of this book. The survey article by Argall et al. [12] presents a more concise review of the field, includes a categorization that highlights differences between approaches, and seeks to identify research areas within LfD that have not yet been explored. Billing and Hellström [32] present a formalism for Learning from Demonstration focused on techniques in which the robot is directly controlled during demonstration (e.g., teleoperation, kinesthetic teaching). The authors present formalisms for concepts such as goal, generalization, and repetition, survey the use of these concepts in the literature and describe them in the light of related concepts in machine learning, planning theory, and psychology. The book *Imitation in Animals and Artifacts* [77] provides an interdisciplinary overview of research in imitation learning, covering leading work from neuroscience, psychology, and linguistics as well as computer science. A narrower focus is presented in the chapter "Robot Programming by Demonstration" [31] within the book *Handbook of Robotics*. This work particularly highlights techniques which may augment or combine with traditional learning methods, such as giving the teacher an active role during learning. Additionally, the first chapter of the book *From Motor Learning to Interaction Learning in Robots* [231] presents a detailed overview of imitation learning from two perspectives, the first coming from research in biology and neurophysiology focused on the cognitive processes required for imitation, and the second reviewing imitation in artificial systems and techniques for learning from demonstrations. Finally, the book *Robot Programming by Demonstration: A Probabilistic Approach* [49] presents a detailed overview of techniques for learning low-level motion trajectories from demonstration.

Bibliography

[1] P. Abbeel, A. Coates, M. Quigley, and A. Y. Ng. An application of reinforcement learning to aerobatic helicopter flight. In *Neural Information Proccessing (NIPS'07)*, 2007. 18, 19, 44

[2] P. Abbeel, A. Coates, and A. Y. Ng. Autonomous Helicopter Aerobatics through Apprenticeship Learning. *The International Journal of Robotics Research*, 29(13):1608–1639, June 2010. DOI: 10.1177/0278364910371999. 26, 44

[3] P. Abbeel and A. Ng. Apprenticeship learning via inverse reinforcement learning. In *Proceedings of the 21st International Conference on Machine Learning*, 2004. DOI: 10.1145/1015330.1015430. 19, 44

[4] D. W. Aha, D. Kibler, and M. K. Albert. Instance-based learning algorithms. *Machine learning*, 6(1):37–66, 1991. DOI: 10.1023/A:1022689900470. 39

[5] B. Akgun, M. Cakmak, J. Yoo, and A. L. Thomaz. Trajectories and Keyframes for Kinesthetic Teaching: A Human-Robot Interaction Perspective. In *Proceedings of theInternational Conference on Human-Robot Interaction*, 2012. DOI: 10.1145/2157689.2157815. 34

[6] E. E. Aksoy, a. Abramov, J. Dorr, K. Ning, B. Dellen, and F. Worgotter. Learning the semantics of object-action relations by observation. *The International Journal of Robotics Research*, 30(10):1229–1249, August 2011. DOI: 10.1177/0278364911410459. 48

[7] R. Aler, O. Garcia, and J. M. Valls. Correcting and improving imitation models of humans for Robosoccer agents. *Evolutionary Computation*, 3(2-5):2402–2409, September 2005. DOI: 10.1109/CEC.2005.1554994. 19

[8] Javier Almingol, Luis Montesano, and Manuel Lopes. Learning multiple behaviors from unlabeled demonstrations in a latent controller space. In *International Conference on Machine Learning*, 2013. 29

[9] Ethem Alpaydin. *Introduction to Machine Learning*. MIT Press, 2004. 38

[10] R Amit and Maja Mataric. Learning movement sequences from demonstration. In *ICDL '02: Proceedings of the 2nd International Conference on Development and Learning*, page 203, Washington, DC, USA, 2002. IEEE Computer Society. DOI: 10.1109/DEVLRN.2002.1011867. 22

[11] B. Argall, B. Browning, and M. Veloso. Learning from Demonstration with the Critique of a Human Teacher. In *Second Annual Conference on Human-Robot Interactions*, Arlington, Virginia, March 2007. DOI: 10.1145/1228716.1228725. 19

[12] B. D. Argall, S. Chernova, M. Veloso, and B. Browning. A survey of robot learning from demonstration. *Robotics and Autonomous Systems*, 57(5):469–483, 2009. DOI: 10.1016/j.robot.2008.10.024. 81

[13] B. D. Argall, E. L. Sauser, and A. G. Billard. Tactile Guidance for Policy Adapatation. pages 79–133, 2010. 10, 19

[14] Brenna Argall and Aude Billard. Learning from Demonstration and Correction via Multiple Modalities for a Humanoid Robot. In *The International Conference SKILLS*, volume 1, 2011. DOI: 10.1051/bioconf/20110100003. 10, 34, 54

[15] Brenna D. Argall. *Learning Mobile Robot Motion Control from Demonstration and Corrective Feedback*. Ph.D. thesis, Robotics Institute, Carnegie Mellon University, Pittsburgh, PA, 2009. 59

[16] A. Aron, E. N. Aron, and E. J. Coups. *Statistics for Psychology, 5th Ed.* Pearson Prentice Hall, New Jersey, 2009. 76, 78

[17] H. Asada. Teaching and learning of compliance using neural nets: representation and generation of nonlinear compliance. In *Robotics and Automation, 1990. Proceedings., 1990 IEEE International Conference on*, pages 1237–1244 vol.2, 1990. DOI: 10.1109/ROBOT.1990.126167. 25, 26

[18] Tamim Asfour, Pedram Azad, Florian Gyarfas, and Rüdiger Dillmann. Imitation learning of dual-arm manipulation tasks in humanoid robots. *International Journal of Humanoid Robotics*, 5(02):183–202, 2008. DOI: 10.1142/S0219843608001431. 32

[19] C. G. Atkeson and S. Schaal. Learning Tasks From A Single Demonstration. In *IEEE International Conference on Robotics and Automation*, number April, 1997. DOI: 10.1109/ROBOT.1997.614389. 43

[20] Christopher G Atkeson and Stefan Schaal. Robot learning from demonstration. In Douglas H. Fisher, Jr., editor, *Proc. 14th International Conference on Machine Learning*, pages 12–20. Morgan Kaufmann, 1997. 28

[21] Monica Babes, Vukosi Marivate, Kaushik Subramanian, and Michael L Littman. Apprenticeship learning about multiple intentions. In *Proceedings of the 28th International Conference on Machine Learning (ICML-11)*, pages 897–904, 2011. 44

[22] C. Bartneck, T. Suzuki, T. Kanda, and T. Nomura. The influence of people's culture and prior experiences with aibo on their attitude towards robots. *AI and Society*, 21(1):217–230, 2007. DOI: 10.1007/s00146-006-0052-7. 70

[23] R. Bellamy. Designing Educational Technology: Computer-Mediated Change. In B Nardi, editor, *Context and Consciousness: Activity Theory and Human-Computer Interaction*. MIT Press, Cambridge, MA, 1996. 11

[24] M. Bennewitz. Learning Motion Patterns of People for Compliant Robot Motion. *The International Journal of Robotics Research*, 24(1):31–48, January 2005. DOI: 10.1177/0278364904048962. 10

[25] D. C. Bentivegna, A. Ude, C. G. Atkeson, and G. Cheng. Humanoid Robot Learning and Game Playing Using PC-Based Vision. In *Proceedings of the IEEE/RSJ International Conference on Intelligent Robots and Systems*, Swiss Federal Institute of Technology Lausanne, Switzerland, October 2002. DOI: 10.1109/IRDS.2002.1041635. 21, 22

[26] M. Berlin, C. Breazeal, and C. Chao. Spatial scaffolding cues for interactive robot learning. In *Proceedings of the23st Conference on Artificial Intelligence*, 2008. DOI: 10.1109/IROS.2008.4651180. 10

[27] C.L. Bethel, J.L. Burke, R.R. Murphy, and K. Salomon. Psychophysiological experimental design for use in human-robot interaction studies. In *Proceedings of the International Symposium on Collaborative Technologies and Systems*, 2007. DOI: 10.1109/CTS.2007.4621744. 78

[28] T. W. Bickmore and R. W. Picard. Towards caring machines. In *CHI'04 Extended Abstracts on Human Factors in Computing Systems*, pages 1489–1492. ACM, 2004. DOI: 10.1145/985921.986097. 70

[29] A. Billard, Y. Epars, G. Cheng, and S. Schaal. Discovering Imitation Strategies through Categorization of Multi-Dimensional Data. In *Proceedings of the 2003 IEEE/RSJ International Conference on Intelligent Robots and Systems*, Las Vegas, Nevada, October 2003. DOI: 10.1109/IROS.2003.1249229. 22

[30] A. Billard and M. Mataric. Learning Human Arm Movements by Imitation: Evaluation of Biologically Inspired Connectionist Architecture. *Robotics and Autonomous Systems*, 37(2-3):145–160, November 2001. DOI: 10.1016/S0921-8890(01)00155-5. 22

[31] Aude Billard, Sylvain Callinon, Rudiger Dillmann, and Stefan Schaal. Robot programming by demonstration. In B Siciliano and O Khatib, editors, *Handbook of Robotics*, chapter 59. Springer, New York, NY, USA, 2008. DOI: 10.1007/978-3-540-30301-5. 81

[32] Erik A Billing and Thomas Hellström. A formalism for learning from demonstration. *Paladyn. Journal of Behavioral Robotics*, 1(1):1–13, 2010. DOI: 10.2478/s13230-010-0001-5. 81

[33] B. Blumberg, M. Downie, Y. Ivanov, M. Berlin, M. P. Johnson, and B. Tomlinson. Integrated learning for interactive synthetic characters. In *Proceedings of the ACM SIGGRAPH*, 2002. DOI: 10.1145/566654.566597. 23, 55

[34] M. Bratman. Shared cooperative activity. *The Philosophical Review*, 101(2):327–341, 1992. DOI: 10.2307/2185537. 12

[35] C Breazeal. *Designing Sociable Robots*. MIT Press, Cambridge, MA, May 2002. 9

[36] C. Breazeal, A. Brooks, J. Gray, G. Hoffman, J. Lieberman, H. Lee, A. Thomaz, and D. Mulanda. Tutelage and Collaboration for Humanoid Robots. *International Journal of Humanoid Robotics*, 1(2), 2004. DOI: 10.1142/S0219843604000150. 20, 43, 59

[37] C. Breazeal, D. Buchsbaum, J. Grey, and B. Blumberg. Learning from and about others: Toward using imitation to bootstrap the social competence of robots. *Artificial Life*, 11, 2005. DOI: 10.1162/1064546053278955. 22

[38] C. Breazeal, N. DePalma, J. Orkin, S. Chernova, and M. Jung. Crowdsourcing human-robot interaction: New methods and system evaluation in a public environment. *Journal of Human-Robot Interaction*, 2(1):82–111, 2013. DOI: 10.5898/JHRI.2.1.Breazeal. 39, 50

[39] C. Breazeal, J. Gray, and M. Berlin. An embodied cognition approach to mindreading skills for socially intelligent robots. *The International Journal of Robotics Research*, 28(5):656–680, May 2009. DOI: 10.1177/0278364909102796. 43

[40] B. Browning, L. Xu, and M. Veloso. Skill acquisition and use for a dynamically-balancing soccer robot. In *Proceedings of Nineteenth National Conference on Artificial Intelligence*, 2004. 19

[41] B. G. Buchanan and T. M. Mitchell. Model-directed learning of production rules. In D. Watermah and F. Hayes-Roth, editors, *Pattern-directed inference systems*. Academic Press, New York, 1978. 43

[42] R. R. Burton, J. S. Brown, and G. Fischer. Skiing as a model of instruction. In B Rogoff and J Lave, editors, *Everyday cognition: its development in social context*. Harvard University Press, Cambridge, MA, 1984. 10

[43] Jesse Butterfield, Sarah Osentoski, Graylin Jay, and Odest Chadwicke Jenkins. Learning from demonstration using a multi-valued function regressor for time-series data. In *Humanoid Robots (Humanoids), 2010 10th IEEE-RAS International Conference on*, pages 328–333. IEEE, 2010. DOI: 10.1109/ICHR.2010.5686284. 50

[44] M. Cakmak. Guided teaching interactions with robots: embodied queries and teaching heuristics. Ph.D. thesis, Georgia Institute of Technology, 2012. 61, 63

[45] M. Cakmak, C. Chao, and A. L. Thomaz. Designing interactions for robot active learners. *IEEE Transactions on Autonomous Mental Development*, 2(2):108–118, 2010. DOI: 10.1109/TAMD.2010.2051030. 54, 60

[46] M. Cakmak, N. DePalma, R. Arriaga, and A. L. Thomaz. Social learning mechanisms for robots. *Autonomous Robots*, 2010. DOI: 10.1007/s10514-010-9197-9. 8

[47] M. Cakmak and A. L. Thomaz. Learning about objects with human teachers. In *Proceedings of theInternational Conference on Human-Robot Interaction*, 2009. DOI: 10.1145/1514095.1514101. 48

[48] M. Cakmak and A. L. Thomaz. Designing robot learners that ask good questions. In *Proceedings of theInternational Conference on Human-Robot Interaction*, 2012. DOI: 10.1145/2157689.2157693. 61, 63

[49] S. Calinon. *Robot Programming by Demonstration: A Probabilistic Approach*. EPFL/CRC Press, 2009. EPFL Press ISBN 978-2-940222-31-5, CRC Press ISBN 978-1-4398-0867-2. 18, 81

[50] S. Calinon and A. Billard. Incremental learning of gestures by imitation in a humanoid robot. In *In Proc. of the ACM/IEEE International Conference on Human-Robot Interaction*, pages 255–262, 2007. DOI: 10.1145/1228716.1228751. 22

[51] S Calinon and A Billard. What is the teacher's role in robot programming by demonstration? - Toward benchmarks for improved learning. *Interaction Studies. Special Issue on Psychological Benchmarks in Human-Robot Interaction*, 8(3), 2007. 18, 20, 34

[52] S. Calinon and A. Billard. A probabilistic programming by demonstration framework handling skill constraints in joint space and task space. In *Proc. IEEE/RSJ International Conference on Intelligent Robots and Systems*, 2008. DOI: 10.1109/IROS.2008.4650593. 34

[53] S. Calinon and A. Billard. Statistical learning by imitation of competing constraints in joint space and task space. *Advanced Robotics*, 23(15):2059–2076, 2009. DOI: 10.1163/016918609X12529294461843. 26, 62

[54] S. Calinon, F. D'halluin, E.L. Sauser, D.G. Caldwell, and A.G. Billard. Learning and reproduction of gestures by imitation. *Robotics Automation Magazine, IEEE*, 17(2):44–54, 2010. DOI: 10.1109/MRA.2010.936947. 34

[55] S. Calinon, F. Guenter, and A. Billard. On learning, representing and generalizing a task in a humanoid robot. *IEEE Transactions on Systems, Man and Cybernetics, Part B. Special*

issue on robot learning by observation, demonstration and imitation, 37(2):286–298, 2007. DOI: 10.1109/TSMCB.2006.886952. 33, 34

[56] J. Call and M. Carpenter. Three sources of information in social learning. In K. Dautenhahn and C.L. Nehaniv, editors, *Imitation in animals and artifacts*. MIT Press, 2002. 8

[57] C. Cazden. Performance before competence: Assistance to child in the ZPD. In M Cole, Y Engeström, and O Vasquez, editors, *Mind, Culture, and Activity: Seminal Papers from the Laboratory of Comparative Human Cognition*. Cambridge University Press, Cambridge, 1997. 11

[58] T. Cederborg, M. Li, A. Baranes, and P.-Y. Oudeyer. Incremental local online gaussian mixture regression for imitation learning of multiple tasks. In *Intelligent Robots and Systems, 2010 IEEE/RSJ International Conference on*, pages 267–274. IEEE, 2010. DOI: 10.1109/IROS.2010.5652040. 46

[59] H. S. Chang. Reinforcement learning with supervision by combining multiple learnings and expert advices. In *Proc. of the American Control Conference*, 2006. DOI: 10.1109/ACC.2006.1657371. 55

[60] C. Chao, M. Cakmak, and A. L. Thomaz. Transparent active learning for robots. In *Autonomous Mental Development, special issue on Active Learning*, 2010. DOI: 10.1145/1734454.1734562. 60

[61] C. Chao, M. Cakmak, and A. L. Thomaz. Towards grounding concepts for transfer in goal learning from demonstration. In *International Conference on Development and Learning*, 2011. DOI: 10.1109/DEVLRN.2011.6037321. 43

[62] C. Chao and A. L. Thomaz. Timing in multimodal turn-taking interactions: Control and analysis using timed Petri nets. *International Journal of Human-Robot Interaction*, 1(1):4–25, 2012. DOI: 10.5898/JHRI.1.1.Chao. 7

[63] C. Chao and A. L. Thomaz. Controlling social dynamics with a parameterized model of floor regulation. *Journal of Human Robot Interaction*, 2013. DOI: 10.5898/JHRI.2.1.Chao. 7

[64] Jason Chen and Alex Zelinsky. Programing by demonstration: Coping with suboptimal teaching actions. *The International Journal of Robotics Research*, 22(5):299–319, May 2003. DOI: 10.1177/0278364903022005002. 19, 34

[65] N. Chentanez, A. G. Barto, and S. P. Singh. Intrinsically motivated reinforcement learning. In *Advances in Neural Information Processing Systems*, pages 1281–1288, 2004. DOI: 10.1007/978-3-642-32375-1_2. 8

[66] S. Chernova and M. Veloso. Interactive Policy Learning through Confidence-Based Autonomy. *Journal of Artificial Intelligence Research*, 34, 2009. DOI: 10.1613/jair.2584. 19, 39, 54, 59, 60, 61

[67] Sonia Chernova and Manuela Veloso. Learning equivalent action choices from demonstration. In *Intelligent Robots and Systems, 2008. IROS 2008. IEEE/RSJ International Conference on*, pages 1216–1221. IEEE, 2008. DOI: 10.1109/IROS.2008.4650995. 50

[68] Silvia Chiappa and Jan Peters. Movement extraction by detecting dynamics switches and repetitions. *Advances in Neural Information Processing Systems*, 23:388–396, 2010. 29

[69] J. Clouse and P. Utgoff. A teaching method for reinforcement learning. In *Proc. of the Nineth International Conference on Machine Learning*, pages 92–101, 1992. 23, 58

[70] J. A. Clouse. On integrating apprentice learning and reinforcement learning. Ph.D. thesis, University of Massachisetts, Department of Computer Science, 1996. 19

[71] Luis C Cobo, Peng Zang, Charles L Isbell Jr, Andrea L Thomaz, and Charles L Isbell Jr. Automatic state abstraction from demonstration. In *Twenty-Second International Joint Conference on Artificial Intelligence*, pages 1243–1248. AAAI Press, 2009. 45

[72] P. R. Cohen, H. J. Levesque, J. H. T. Nunes, and S. L. Oviatt. Task-oriented dialogue as a consequence of joint activity. In *Proceedings of thePacific Rim International Conference on Artificial Intelligence*, Nagoya, Japan, November 1990. 11

[73] D. A. Cohn, Z. Ghahramani, and M. I. Jordan. Active learning with statistical models. *Journal of Artificial Intelligence Research*, 4:129–145, 1996. 60

[74] K. Conn and R. A. Peters. Reinforcement learning with a supervisor for a mobile robot in a real-world environment. In *Computational Intelligence in Robotics and Automation, 2007. CIRA 2007. International Symposium on*, pages 73–78. IEEE, 2007. DOI: 10.1109/CIRA.2007.382878. 58

[75] C Crick, S Osentoski, G Jay, and O Jenkins. Human and robot perception in large-scale learning from demonstration. In *ACM/IEEE International Conference on Human-Robot Interaction (HRI 2011)*, 2011. DOI: 10.1145/1957656.1957788. 38

[76] G. Csibra. Teleological and referential understanding of action in infancy. *Phil. Trans. The Royal Society of London*, 358:447–458, 2003. DOI: 10.1098/rstb.2002.1235. 10

[77] K. Dautenhahn and C. L. Nehaniv, editors. *Imitation in animals and artifacts*. MIT Press, Cambridge, MA, USA, 2002. 81

[78] K. Dautenhahn and I. Werry. A quantitative technique for analysing robot-human interactions. In *Intelligent Robots and Systems, 2002. IEEE/RSJ International Conference on*, volume 2, pages 1132–1138 vol.2, 2002. DOI: 10.1109/IRDS.2002.1043883. 71

[79] M. P. Deisenroth, C. E. Rasmussen, and D. Fox. Learning to control a low-cost manipulator using data-efficient reinforcement learning. In *Robotics: Science & Systems*, 2011. 22

[80] J. Demiris and G. Hayes. Imitation as a dual-route process featuring predictive and learning components: a biologically plausible computational model. In K Dautenhahn and C L Nehaniv, editors, *Imitation in Animals and Artifacts*. MIT Press, Cambridge, 2002. 22

[81] A. P. Dempster, N. M. Laird, and D. B. Rubin. Maximum likelihood from incomplete data via the em algorithm. *Journal of the Royal Statistical Society. Series B (Methodological)*, pages 1–38, 1977. DOI: 10.2307/2984875. 31

[82] R Dillmann. Teaching and learning of robot tasks via observation of human performance. *Robotics and Autonomous Systems*, 47(2-3):109–116, 2004. DOI: 10.1016/j.robot.2004.03.005. 41

[83] K. R. Dixon. Predictive robot programming: Theoretical and experimental analysis. *The International Journal of Robotics Research*, 23(9):955–973, September 2004. DOI: 10.1177/0278364904044401. 34

[84] Shuonan Dong and Brian Williams. Motion learning in variable environments using probabilistic flow tubes. In *Robotics and Automation , 2011 IEEE International Conference on*, pages 1976–1981. IEEE, 2011. DOI: 10.1109/ICRA.2011.5980530. 46

[85] Shuonan Dong and Brian Williams. Learning and recognition of hybrid manipulation motions in variable environments using probabilistic flow tubes. *International Journal of Social Robotics*, 4(4):357–368, 2012. DOI: 10.1007/s12369-012-0155-x. 46, 47

[86] John V Draper and Linda M Blair. Workload, flow, and telepresence during teleoperation. In *Robotics and Automation, 1996. Proceedings. 1996 IEEE International Conference on*, volume 2, pages 1030–1035. IEEE, 1996. DOI: 10.1109/ROBOT.1996.506844. 70

[87] G. Druck, B. Settles, and A. McCallum. Active learning by labeling features. In *In Proceedings of the Conference on Empirical Methods in Natural Language Processing*, pages 81–90, 2009. 62

[88] B. Dufay and J.-C. Latombe. An approach to automatic robot programming based on inductive learning. *The International Journal of Robotics Research*, 3(4):3–20, 1984. DOI: 10.1177/027836498400300401. 26

[89] Bruno Dufay and Jean-Claude Latombe. An approach to automatic robot programming based on inductive learning. *The International Journal of Robotics Research*, 3(4):3–20, 1984. DOI: 10.1177/027836498400300401. 25

[90] M Ehrenmann, R Zollner, O Rogalla, and R Dillmann. Programming service tasks in household environments by human demonstration. In *Robot and Human Interactive Communication*, 2002. Proceedings 11th IEEE International Workshop on, pages 460–467. IEEE, 2002. DOI: 10.1109/ROMAN.2002.1045665. 41

[91] R. Evans. Varieties of Learning. In S Rabin, editor, *AI Game Programming Wisdom*, pages 567–578. Charles River Media, Hingham, MA, 2002. 23

[92] P. Fitzpatrick, G. Metta, L. Natale, S. Rao, and G. Sandini. Learning about objects through action-initial steps towards artificial cognition. In *Robotics and Automation, 2003. Proceedings. ICRA'03. IEEE International Conference on*, volume 3, pages 3140–3145. IEEE, 2003. DOI: 10.1109/ROBOT.2003.1242073. 47

[93] Terrence Fong, Illah Nourbakhsh, Robert Ambrose, Reid Simmons, Alan Schultz, and Jean Scholtz. The peer-to-peer human-robot interaction project. In *AIAA Space*, volume 2005, 2005. 70

[94] H. Friedrich, S. Münch, R. Dillmann, S. Bocionek, and M. Sassin. Robot programming by demonstration (rpd): Supporting the induction by human interaction. *Machine Learning*, 23(2-3):163–189, 1996. DOI: 10.1007/BF00117443. 43

[95] A. Garland and N. Lesh. Learning hierarchical task models by demonstration. Technical Report TR2003-01, Mitsubishi Electric Research Laboratories, 2003. 40

[96] J. J. Gibson. *The Ecological Approach to Visual Perception*. Routledge, 1986. 47

[97] A. Gopnik, D. Sobel, L. Schulz, and C. Glymour. Causal learning mechanisms in very young children: Two, three, and four-year-olds infer causal relations from patterns of variation and covariation. *Developmental Psychology*, 37(5):620–629, 2001. DOI: 10.1037/0012-1649.37.5.620. 12

[98] A. C. Graesser and N. K. Person. Question asking during tutoring. *American Educational Research Journal*, 31(1):104–137, 1994. DOI: 10.3102/00028312031001104. 12

[99] P. M. Greenfield. Theory of the teacher in learning activities of everyday life. In B Rogoff and J Lave, editors, *Everyday cognition: its development in social context*. Harvard University Press, Cambridge, MA, 1984. 7, 9

[100] E Gribovskaya, Florent D'Halluin, and Aude Billard. An active learning interface for bootstrapping robot's generalization abilities in learning from demonstration. In *RSS Workshop Towards Closing the Loop: Active Learning for Robotics*, 2010. 62

[101] D. H. Grollman and A. G. Billard. Learning from failure. In *Proceedings of the 6th International Conference on Human-Robot Interaction*, pages 145–146. ACM, 2011. 62

[102] D. H. Grollman and O. C. Jenkins. Dogged learning for robots. In *Proceedings of the IEEE International Conference on Robotics and Automation*, Roma, Italy, April 2007. DOI: 10.1109/ROBOT.2007.363692. 19, 24, 39, 54, 60

[103] Daniel H. Grollman and Aude G. Billard. Robot learning from failed demonstrations. *International Journal of Social Robotics*, 4(4):331–342, June 2012. DOI: 10.1007/s12369-012-0161-z. 35

[104] G. Z. Grudic and P. D. Lawrence. Human-to-robot skill transfer using the SPORE approximation. In *Proceedings of the IEEE International Conference on Robotics and Automation*, Minneapolis, Minnnesota, April 1996. DOI: 10.1109/ROBOT.1996.509162. 24

[105] F. Guenter and A. G. Billard. Using reinforcement learning to adapt an imitation task. In *IEEE/RSJ International Conference on Intelligent Robots and Systems*, 2007. DOI: 10.1109/IROS.2007.4399449. 44

[106] Isabelle Guyon and André Elisseeff. An introduction to variable and feature selection. *The Journal of Machine Learning Research*, 3:1157–1182, 2003. 45

[107] P. Hakkarainen. Play and motivation. In Y. Engeström, R. Miettinen, and R-L. Punamäki, editors, *Perspectives on Activity Theory*. Cambridge University Press, Cambridge, MA, 1996. 7

[108] Sandra G Hart and Lowell E Staveland. Development of nasa-tlx (task load index): Results of empirical and theoretical research. *Human Mental Workload*, 1(3):139–183, 1988. DOI: 10.1016/S0166-4115(08)62386-9. 70

[109] G. Hayes and J. Demiris. A Robot Controller Using Learning by Imitation. In *Proceedings of the 2nd International Symposium on Intelligent Robotic Systems*, Grenoble, France, July 1994. 20

[110] A. Holroyd, C. Rich, C. Sidner, and B. Ponsler. Generating connection events for human-robot collaboration. In *Proceedings of the IEEE International Symposium on Robot and Human Interactive Communication*, pages 241–246, 2011. DOI: 10.1109/RO-MAN.2011.6005245. 7

[111] K. Hornbaek. Some whys and hows of experiments in human-computer interaction. *Foundations and Trends in Human-Computer Interaction*, 5(4):299–373, 2011. DOI: 10.1561/1100000043. 78

[112] A. O. Horvath and L. S. Greenberg. Development and validation of the working alliance inventory. *Journal of Counseling Psychology*, 36(2):223, 1989. DOI: 10.1037/0022-0167.36.2.223. 70

[113] G. Hovland, P. Sikka, and B. McCarragher. Skill acquisition from human demonstration using a hidden markov model. In *IEEE International Conference on Robotics and Automation*, 1996. DOI: 10.1109/ROBOT.1996.506571. 26

[114] A. J. Ijspeert, J. Nakanishi, and S. Schaal. Learning rhythmic movements by demonstration using nonlinear oscillators. In *Proceedings of the IEEE/RSJ Int. Conference on Intelligent Robots and Systems*, 2002. DOI: 10.1109/IRDS.2002.1041514. 21, 27

[115] A. J. Ijspeert, J. Nakanishi, and S. Schaal. Movement imitation with nonlinear dynamical systems in humanoid robots. In *Proceedings of the IEEE International Conference on Robotics and Automation*, 2002. DOI: 10.1109/ROBOT.2002.1014739. 21

[116] Katsushi Ikeuchi. Assembly plan from observation. In *Electronic Manufacturing Technology Symposium, 1995, Proceedings of 1995 Japan International, 18th IEEE/CPMT International*, pages 9–12. IEEE, 1995. 40

[117] M. Imai, T. Kanda, T. Ono, H. Ishiguro, and K. Mase. Robot mediated round table: Analysis of the effect of robot's gaze. In *Proceedings. 11th IEEE International Workshop on Robot and Human Interactive Communication*, pages 411–416, 2002. DOI: 10.1109/RO-MAN.2002.1045657. 10

[118] T. Inamura, M. Inaba, and H. Inoue. Acquisition of probabilistic behavior decision model based on the interactive teaching method. In *Proceedings of the Ninth International Conference on Advanced Robotics*, pages 523–528, 1999. 19, 39, 59

[119] C. Isbell, C. Shelton, M. Kearns, S. Singh, and P. Stone. Cobot: A social reinforcement learning agent. *5th International Conference on Autonomous Agents*, 2001. 23

[120] C. L. Isbell, C. Shelton, M. Kearns, S. Singh, and P. Stone. A social reinforcement learning agent. In *Proc. of the 19th AAAI Conference on Artificial Intelligence*, pages 377–384, 2001. 55

[121] D. Blank J. B. Marshall and L. Meeden. An emergent framework for self-motivation in developmental robotics. In *International Conference on Development and Learning*, pages 104–111, 2004. 8

[122] O. C. Jenkins and M. Matarić. Deriving action and behavior primitives from human motion data. In *Proceedings of the IEEE/RSJ International Conference on Intelligent Robots and Systems*, pages 2551–2556, 2002. DOI: 10.1109/IRDS.2002.1041654. 22, 26

[123] Odest Chadwicke Jenkins and Maja J Matarić. Performance-derived behavior vocabularies: Data-driven acquisition of skills from motion. *International Journal of Humanoid Robotics*, 2004. DOI: 10.1142/S0219843604000186. 22

[124] Nikolay Jetchev and Marc Toussaint. Task space retrieval using inverse feedback control. In *Proceedings of the 28th International Conference on Machine Learning (ICML-11)*, pages 449–456, 2011. 63

[125] Leslie Pack Kaelbling, Michael L Littman, and Andrew W Moore. Reinforcement learning: A survey. *arXiv preprint cs/9605103*, 1996. 55

[126] M. Kaiser, H. Friedrich, and R. Dillmann. Obtaining good performance from a bad teacher. In *Programming by Demonstration vs. Learning from Examples Workshop at ML'95*, 1995. 34

[127] T. Kanda, H. Ishiguro, M. Imai, and T. Ono. Body movement analysis of human-robot interaction. pages 177–182, August 2003. 10

[128] T. Kanda, H. Ishiguro, M. Imai, and T. Ono. Development and evaluation of interactive humanoid robots. In *Proceedings of the IEEE*, volume 92, pages 1839–1850, 2004. DOI: 10.1109/JPROC.2004.835359. 7

[129] F. Kaplan, P.-Y. Oudeyer, E. Kubinyi, and A. Miklosi. Robotic clicker training. *Robotics and Autonomous Systems*, 38(3-4):197–206, 2002. DOI: 10.1016/S0921-8890(02)00168-9. 10, 23, 55

[130] K. Kaye. Infant's effects upon their mothers' teaching strategies. In J Glidewell, editor, *The Social Context of Learning and Development*. Gardner Press, New York, 1977. 7

[131] S. M. Khansari-Zadeh and A. Billard. Learning stable non-linear dynamical systems with Gaussian mixture models. *IEEE Transaction on Robotics*, 2011. DOI: 10.1109/TRO.2011.2159412. 34

[132] C. D. Kidd and C. Breazeal. Robots at home: Understanding long-term human-robot interaction. In *Intelligent Robots and Systems*, pages 3230–3235. IEEE, 2008. DOI: 10.1109/IROS.2008.4651113. 70

[133] Hedvig Kjellström, Javier Romero, and Danica Kragić. Visual object-action recognition: Inferring object affordances from human demonstration. *Computer Vision and Image Understanding*, 115(1):81–90, 2011. DOI: 10.1016/j.cviu.2010.08.002. 48, 49

[134] W. B. Knox and P. Stone. Tamer: Training an agent manually via evaluative reinforcement. In *Proc. of the 7th IEEE International Conference on Development and Learning*, pages 292–297, 2008. 55, 56

[135] W. B. Knox and P. Stone. Interactively shaping agents via human reinforcement: The tamer framework. In *The Fifth International Conference on Knowledge Capture*, September 2009. DOI: 10.1145/1597735.1597738. 23, 56

[136] W. B. Knox and P. Stone. Combining manual feedback with subsequent mdp reward signals for reinforcement learning. In *Proc. of the 9th International Conference on Autonomous Agents and Multiagent Systems*, pages 5–12, 2010. DOI: 10.1145/1838206.1838208. 58

[137] W. B. Knox and P. Stone. Reinforcement learning from simultaneous human and mdp reward. In *Proc. of the 11th International Conference on Autonomous Agents and Multiagent Systems*, pages 475–482, 2012. 58

[138] W. B. Knox, P. Stone, and C. Breazeal. Training a robot via human feedback: A case study. In *International Conference on Social Robotics*, pages 460–470. Springer, 2013. DOI: 10.1007/978-3-319-02675-6_46. 23, 56

[139] J. Kober, J. A. Bagnell, and J. Peters. Reinforcement learning in robotics: A survey. *The International Journal of Robotics Research*, 32(11):1238–1274, 2013. DOI: 10.1177/0278364913495721. 59

[140] J. Kober and J. Peters. Reinforcement learning in robotics: A survey. In *Reinforcement Learning*, pages 579–610. Springer, 2012. DOI: 10.1007/978-3-642-27645-3_18. 53, 55, 59

[141] J. Koenemann, F. Burget, and M. Bennewitz. Real-time imitation of human whole-body motions by humanoids. In *Proc. of the IEEE International Conference on Robotics & Automation (ICRA)*, 2014. 19, 20

[142] Jens Kohlmorgen and Steven Lemm. An on-line method for segmentation and identification of non-stationary time series. In *Neural Networks for Signal Processing XI, 2001. Proceedings of the 2001 IEEE Signal Processing Society Workshop*, pages 113–122. IEEE, 2001. DOI: 10.1109/NNSP.2001.943116. 31

[143] J Zico Kolter, Pieter Abbeel, and Andrew Y Ng. Hierarchical apprenticeship learning with application to quadruped locomotion. *Advances in Neural Information Processing Systems*, 20:769–776, 2008. 21, 44

[144] George Konidaris, Scott Kuindersma, Roderic Grupen, and Andrew Barto. Robot learning from demonstration by constructing skill trees. *The International Journal of Robotics Research*, 31(3):360–375, December 2011. DOI: 10.1177/0278364911428653. 44

[145] P Kormushev, S Calinon, and D G Caldwell. Robot Motor Skill Coordination with EM-based Reinforcement Learning. In *Proc. IEEE/RSJ International Conference on Intelligent Robots and Systems*, 2010. DOI: 10.1109/IROS.2010.5649089. 26

[146] Sotiris B Kotsiantis, ID Zaharakis, and PE Pintelas. *Supervised machine learning: A review of classification techniques*. 2007. 38

[147] V. Kruger, D. Herzog, S. Baby, A. Ude, and D. Kragic. Learning actions from observations. *Robotics & Automation Magazine, IEEE*, 17(2):30–43, 2010. DOI: 10.1109/MRA.2010.936961. 32

[148] G. Kuhlmann, P. Stone, R. J. Mooney, and J. W. Shavlik. Guiding a reinforcement learner with natural language advice: Initial results in roboCup soccer. In *Proceedings of theAAAI-2004 Workshop on Supervisory Control of Learning and Adaptive Systems*, San Jose, CA, July 2004. 23, 58

[149] D. Kulic, C. Ott, D. Lee, J. Ishikawa, and Y. Nakamura. Incremental learning of full body motion primitives and their sequencing through human motion observation. *The International Journal of Robotics Research*, 31(3):330–345, November 2011. DOI: 10.1109/ICHR.2008.4756000. 31, 32

[150] D. Kulic, W. Takano, and Y. Nakamura. Incremental learning, clustering and hierarchy formation of whole body motion patterns using adaptive hidden markov chains. *International Journal of Robotics Research*, 27(7):761–784, July 2008. DOI: 10.1177/0278364908091153. 31, 32

[151] Dana Kulic, Wataru Takano, and Yoshihiko Nakamura. Online segmentation and clustering from continuous observation of whole body motions. *IEEE Transactions on Robotics*, 25(5):641–647, 2009. DOI: 10.1109/TRO.2009.2026508. 31, 32

[152] Yasuo Kuniyoshi, Masayuki Inaba, and Hirochika Inoue. Learning by watching: Extracting reusable task knowledge from visual observation of human performance. In *IEEE Transactions on Robotics and Automation*, volume 10, pages 799–822, 1994. DOI: 10.1109/70.338535. 22, 40

[153] Yoshinori Kuno, Kazuhisa Sadazuka, Michie Kawashima, Keiichi Yamazaki, Akiko Yamazaki, and Hideaki Kuzuoka. Museum guide robot based on sociological interaction analysis. In *Proceedings of the SIGCHI Conference on Human Factors in Computing Systems - CHI '07*, page 1191, New York, New York, USA, April 2007. ACM Press. DOI: 10.1145/1240624.1240804. 10

[154] John Lafferty, Andrew McCallum, and Fernando CN Pereira. Conditional random fields: Probabilistic models for segmenting and labeling sequence data. 2001. 49

[155] J. Lave and E. Wenger. *Situated Learning: Legitimate Peripheral Participation*. Cambridge University Press, Cambridge, 1991. DOI: 10.1017/CBO9780511815355. 8

[156] D. Lee and Y. Nakamura. Mimesis model from partial observations for a humanoid robot. *The International Journal of Robotics Research*, 29(1):60–80, August 2009. DOI: 10.1177/0278364909342282. 32

[157] D. Lee, C. Ott, and Y. Nakamura. Mimetic communication model with compliant physical contact in human–humanoid interaction. *The International Journal of Robotics Research*, 29(13):1684–1704, May 2010. DOI: 10.1177/0278364910364164. 32

[158] S. Levine, Z. Popovic, and V. Koltun. Feature construction for inverse reinforcement learning. In *Advances in Neural Information Processing Systems*, pages 1342–1350, 2010. 45

[159] J. Lieberman and C. Breazeal. Improvements on action parsing and action interpolation for learning through demonstration. In *4th IEEE/RAS International Conference on Humanoid Robots*, volume 1, pages 342–365, 2004. DOI: 10.1109/ICHR.2004.1442131. 19

[160] R. Likert. A technique for the measurement of attitudes. *Archives of Psychology*, 140:1–55, 1932. 69

[161] M. Likhachev and R. C. Arkin. Spatio-temporal case-based reasoning for behavioral selection. In *Robotics and Automation, 2001. Proceedings 2001 ICRA. IEEE International Conference on*, volume 2, pages 1627–1634. IEEE, 2001. DOI: 10.1109/ROBOT.2001.932844. 39

[162] Y. Lin, S. Ren, M. Clevenger, and Y. Sun. Learning grasping force from demonstration. In *Robotics and Automation , 2012 IEEE International Conference on*, pages 1526–1531, 2012. DOI: 10.1109/ICRA.2012.6225222. 26

[163] B. Litowitz. Just say no: Responsibility and resistance. In M Cole, Y Engeström, and O Vasquez, editors, *Mind, Culture, and Activity: Seminal Papers from the Laboratory of Comparative Human Cognition*. Cambridge University Press, Cambridge, 1997. 8

[164] Huan Liu and Hiroshi Motoda. *Feature Selection for Knowledge Discovery and Data Mining*. Springer, 1998. DOI: 10.1007/978-1-4615-5689-3. 45

[165] A. Lockerd and C. Breazeal. Tutelage and socially guided robot learning. In *IEEE/RSJ International Conference on Intelligent Robots and Systems*, 2004. DOI: 10.1109/IROS.2004.1389954. 39, 59

[166] R. Lomasky, C. Brodley, M. Aernecke, D. Walt, and M. Friedl. Active class selection. In *Machine Learning: ECML 2007*, volume 4701 of *Lecture Notes in Computer Science*, pages 640–647. Springer-Verlag, 2007. DOI: 10.1007/978-3-540-74958-5_63. 62

[167] M. Lopes, F. Melo, and L. Montesano. Active learning for reward estimation in inverse reinforcement learning. In *Machine Learning and Knowledge Discovery in Databases*, pages 31–46. Springer Berlin/Heidelberg, 2009. DOI: 10.1007/978-3-642-04174-7_3. 44, 61

[168] Manuel Lopes, Francisco Melo, Luis Montesano, and Jose Santon-Victor. Abstraction levels for robotic imitation: Overview and computational approaches. In *From Motor*

Learning to Interaction Learning in Robots, pages 313–355. 2010. DOI: 10.1007/978-3-642-05181-4_14. 48

[169] Manuel Lopes, Francisco S. Melo, and Luis Montesano. Affordance-based imitation learning in robots. *2007 IEEE/RSJ International Conference on Intelligent Robots and Systems*, pages 1015–1021, October 2007. DOI: 10.1109/IROS.2007.4399517. 48

[170] R. Maclin, J. Shavlik, L. Torrey, T. Walker, and E. Wild. Giving advice about preferred actions to reinforcement learners via knowledge-based kernel regression. In *Proceedings of theThe Twentieth National Conference on Artificial Intelligence*, Pittsburgh, PA, July 2005. 23, 58

[171] Y. Matsusaka, S. Fujie, and T. Kobayashi. Modeling of conversational strategy for the robot participating in the group conversation. In *INTERSPEECH'01*, pages 2173–2176, 2001. 10

[172] B. J. McCarragher and H. Asada. The discrete event modeling and trajectory planning of robotic assembly tasks. *Journal of Dynamic Systems, Measurement, and Control*, 117:394–394, 1995. DOI: 10.1115/1.2799130. 25

[173] Wim Meeussen, Johan Rutgeerts, Klaas Gadeyne, Herman Bruyninckx, and Joris De Schutter. Contact-state segmentation using particle filters for programming by human demonstration in compliant-motion tasks. *IEEE Transactions on Robotics*, 23(2):218–231, April 2007. DOI: 10.1109/TRO.2007.892227. 26

[174] A. N. Meltzoff. The Human infant as imitative generalist: A 20-year progress report on infant imitation with implications for comparative psychology. In B G Galef, C. M. Heyes, editor, *Social Learning in Animals: The Roots of Culture*. Academic Press, San Diego, CA, 1996. 7

[175] Çetin Meriçli, Steven D Klee, Jack Paparian, and Manuela Veloso. An interactive approach for situated task teaching through verbal instructions. 2013. 20, 59

[176] Çetin Meriçli, Manuela Veloso, and H Levent Akın. Multi-resolution corrective demonstration for efficient task execution and refinement. *International Journal of Social Robotics*, 4(4):423–435, 2012. DOI: 10.1007/s12369-012-0159-6. 45

[177] L. Montesano, M. Lopes, A. Bernardino, and J. Santos-Victor. Learning object affordances: From sensory-motor coordination to imitation. *IEEE Transactions on Robotics*, 24:15–26, 2008. DOI: 10.1109/TRO.2007.914848. 48

[178] J. R. Movellan. An infomax controller for real time detection of social contingency. In *Proceedings of the 4th International Conference on Development and Learning*, volume 1, pages 19–24, 2005. DOI: 10.1109/DEVLRN.2005.1490937. 12

[179] M Muhlig, Michael Gienger, Jochen J Steil, and Christian Goerick. Automatic selection of task spaces for imitation learning. In *Intelligent Robots and Systems, 2009. IROS 2009. IEEE/RSJ International Conference on*, pages 4996–5002. IEEE, 2009. DOI: 10.1109/IROS.2009.5353894. 34, 63

[180] K. Mulling, J. Kober, O. Kroemer, and J. Peters. Learning to select and generalize striking movements in robot table tennis. *The International Journal of Robotics Research*, 32(3):263–279, January 2013. DOI: 10.1177/0278364912472380. 26, 29

[181] B. Mutlu, J. Forlizzi, and J. Hodgins. A storytelling robot: Modeling and evaluation of human-like gaze behavior. In *2006 6th IEEE-RAS International Conference on Humanoid Robots*, pages 518–523. IEEE, December 2006. DOI: 10.1109/ICHR.2006.321322. 10

[182] B. Mutlu, T. Shiwa, T. Kanda, H. Ishiguro, and N. Hagita. Footing in human-robot conversations: how robots might shape participant roles using gaze cues. In *Proceedings of the 2009 ACM Conference on Human-Robot Interaction*, 2009. DOI: 10.1145/1514095.1514109. 10

[183] Y. Nagai, C. Muhl, and K. J. Rohlfing. Toward designing a robot that learns actions from parental demonstrations. In *Proceedings of the 2008 IEEE International Conference on Robotics and Automation*, pages 3545–3550, 2008. DOI: 10.1109/ROBOT.2008.4543753. 10

[184] Jun Nakanishi, Jun Morimoto, Gen Endo, Gordon Cheng, Stefan Schaal, and Mitsuo Kawato. Learning from demonstration and adaptation of biped locomotion. *Robotics and Autonomous Systems*, 47:79–91, 2004. DOI: 10.1016/S0921-8890(04)00039-9. 21

[185] M. Nakano, Y. Hasegawa, K. Funakoshi, J. Takeuchi, T. Torii, K. Nakadai, N. Kanda, K. Komatani, H. G. Okuno, and H. Tsujino. A multi-expert model for dialogue and behavior control of conversational robots and agents. *Knowledge-Based Systems*, 24(2):248–256, 2011. DOI: 10.1016/j.knosys.2010.08.004. 7

[186] Chrystopher L Nehaniv and Kerstin Dautenhahn. The correspondence problem. *Imitation in Animals and Artifacts*, pages 41–61, 2002. 18

[187] U. Nehmzow, O. Akanyeti, C. Weinrich, T. Kyriacou, and S. A. Billings. Robot programming by demonstration through system identification. In *IEEE/RSJ International Conference on Intelligent Robots and Systems*, 2007. DOI: 10.1109/IROS.2007.4399087. 20

[188] Gergely Neu and Csaba Szepesvári. Apprenticeship learning using inverse reinforcement learning and gradient methods. *arXiv preprint arXiv:1206.5264*, 2012. 44

[189] A. Y. Ng and S. J. Russell. Algorithms for inverse reinforcement learning. In *Icml*, pages 663–670, 2000. 44

[190] M. N. Nicolescu and M. J. Matarić. Experience-based representation construction: learning from human and robot teachers. In *Proceedings of the IEEE/RSJ International Conference on Intelligent Robots and Systems*, Maui, Hawaii, 2001. DOI: 10.1109/IROS.2001.976257. 20

[191] M. N. Nicolescu and M. J. Matarić. Methods for robot task learning: Demonstrations, generalization and practice. In *Second International Joint Conference on Autonomous Agents and Multi-Agent Systems*, Melbourne, Australia, July 2003. DOI: 10.1145/860575.860614. 40, 59

[192] Scott Niekum, Sachin Chitta, Bhaskara Marthi, Sarah Osentoski, and Andrew G Barto. Incremental semantically grounded learning from demonstration. In *Robotics Science and Systems*, 2013. 41

[193] Scott Niekum, Sarah Osentoski, George Konidaris, and Andrew G Barto. Learning and generalization of complex tasks from unstructured demonstrations. In *Intelligent Robots and Systems, 2012 IEEE/RSJ International Conference on*, pages 5239–5246. IEEE, 2012. DOI: 10.1109/IROS.2012.6386006. 29, 41, 46

[194] Scott D Niekum. Semantically grounded learning from unstructured demonstrations. Ph.D. thesis, University of Massachusetts Amherst, 2013. 41, 42

[195] T. Nomura, T. Suzuki, T. Kanda, and K. Kato. Measurement of negative attitudes toward robots. *Interaction Studies*, 7(3):437–454, 2006. DOI: 10.1075/is.7.3.14nom. 70

[196] M. Ollis, W. H. Huang, and M. Happold. A Bayesian approach to imitation learning for robot navigation. In *IEEE/RSJ International Conference on Intelligent Robots and Systems*, 2007. DOI: 10.1109/IROS.2007.4399220. 45

[197] D. R. Olsen and M. A. Goodrich. Metrics for evaluating human-robot interactions. In *Proceedings of the Performance Metrics for Intelligent Systems Workshop (PerMIS)*, 2003. 78

[198] S. Ontañón, K. Mishra, N. Sugandh, and A. Ram. Case-based planning and execution for real-time strategy games. In *Case-Based Reasoning Research and Development*, pages 164–178. Springer, 2007. DOI: 10.1007/978-3-540-74141-1_12. 39

[199] C. C. Graylin, J. Dong, S. Grollman, D. Suay, H.B. Osentoski, S. Pitzer, B. Jenkins, O.C., Sarah Osentoski, Benjamin Pitzer, Christopher Crick, Graylin Jay, Shuonan Dong, Daniel Grollman, Halit Bener Suay, and Odest Chadwicke Jenkins. Remote robotic laboratories for learning from demonstration. *International Journal of Social Robotics*, pages 1–13, 2012. DOI: 10.1007/s12369-012-0157-8. 79

[200] Pierre-yves Oudeyer, Adrien Baranes, and Frédéric Kaplan. Intrinsically motivated exploration for developmental and active sensorimotor learning. In *From Motor Learning to Interaction Learning in Robots*, pages 107–146. 2010. DOI: 10.1007/978-3-642-05181-4_6. 8

[201] Peter Pastor, Heiko Hoffmann, Tamim Asfour, and Stefan Schaal. Learning and generalization of motor skills by learning from demonstration. In *IEEE International Conference on Robotics and Automation*, pages 763–768. IEEE, 2009. DOI: 10.1109/ROBOT.2009.5152385. 28

[202] Peter Pastor, Mrinal Kalakrishnan, Sachin Chitta, Evangelos Theodorou, and Stefan Schaal. Skill learning and task outcome prediction for manipulation. *IEEE International Conference on Robotics and Automation*, pages 3828–3834, May 2011. DOI: 10.1109/ICRA.2011.5980200. 26, 28, 29, 53

[203] R. Pea. Practices of distributed intelligence and designs for education. In G Salomon, editor, *Distributed Cognitions: Psychological and Educational Considerations*. Cambridge University Press, New York, 1993. 7

[204] N. K. Person, A. C. Graesser, J. P. Magliano, and R. J. Kreuz. Inferring what the student knows in one-to-one tutoring: the role of student questions and answers. *Learning and Individual Differences*, 6(2):205–229, 1994. DOI: 10.1016/1041-6080(94)90010-8. 12

[205] Jan Peters and Stefan Schaal. Reinforcement learning of motor skills with policy gradients. *Neural networks: The Official Journal of the International Neural Network Society*, 21(4):682–697, May 2008. DOI: 10.1016/j.neunet.2008.02.003. 29

[206] Thair Nu Phyu. Survey of classification techniques in data mining. In *Proceedings of the International MultiConference of Engineers and Computer Scientists*, volume 1, pages 18–20, 2009. 38

[207] J. Piaget. *The Origins of Intelligence in Children*. International Universities Press, 1952. DOI: 10.1037/11494-000. 12

[208] P. M. Pilarski, M. R. Dawson, T. Degris, F. Fahimi, J. P. Carey, and R. S. Sutton. Online human training of a myoelectric prosthesis controller via actor-critic reinforcement learning. In *Proc. of the IEEE International Conference on Rehabilitation Robotics*, pages 1–7, 2011. 55

[209] N. Pollard and J. K. Hodgins. Generalizing demonstrated manipulation tasks. In *Workshop on the Algorithmic Foundations of Robotics*, December 2002. DOI: 10.1007/978-3-540-45058-0_31. 22

[210] John Ross Quinlan. *C4. 5: Programs for Machine Learning*, volume 1. Morgan Kaufmann, 1993. 38

[211] L. R. Rabiner. A tutorial on hidden markov models and selected applications in speech recognition. *Proceedings of the IEEE*, 77(2):257–286, 1989. DOI: 10.1109/5.18626. 30

[212] H. Raghavan, O. Madani, and R. Jones. Active learning with feedback on features and instances. *Journal of Machine Learning Research*, 7:1655–1686, 2006. 62

[213] D. Ramachandran and E. Amir. Bayesian inverse reinforcement learning. In *International Joint Conference on Artificial Intelligence*, pages 2586–2591, 2007. 44

[214] Rajesh P N Rao, Aaron P Shon, and Andrew N Meltzoff. A Bayesian model of imitation in infants and robots. *Imitation and Social Learning in Robots, Humans, and Animals: Behavioural, Social and Communicative Dimensions*, 2004. 19, 39

[215] N. Ratliff, J. A. Bagnell, and M. A. Zinkevich. Maximum margin planning. In *Proceedings of the 23rd International Conference on Machine Learning*, Pittsburgh, Pennsylvannia, 2006. DOI: 10.1145/1143844.1143936. 21

[216] N. Ratliff, D. Bradley, J. A. Bagnell, and J. Chestnutt. Boosting structured prediction for imitation learning. *Advances in Neural Information Processing Systems 19*, 2007. 21

[217] Nathan D Ratliff, David Silver, and J Andrew Bagnell. Learning to search: Functional gradient techniques for imitation learning. *Autonomous Robots*, 27(1):25–53, 2009. DOI: 10.1007/s10514-009-9121-3. 44, 79

[218] C. Rich, B. Ponsler, A. Holroyd, and C. L. Sidner. Recognizing engagement in human-robot interaction. In *Proceedings of the 2010 ACM Conference on Human-Robot Interaction*, 2010. DOI: 10.1109/HRI.2010.5453163. 7

[219] B Rogoff and H Gardner. Adult guidance of cognitive development. In B Rogoff and J Lave, editors, *Everyday Cognition: Its Development in Social Context*. Harvard University Press, Cambridge, MA, 1984. 7, 10, 11

[220] R. Ros, R. L. De Màntaras, J. L. Arcos, and M. Veloso. Team playing behavior in robot soccer: A case-based reasoning approach. In *Case-Based Reasoning Research and Development*, pages 46–60. Springer, 2007. DOI: 10.1007/978-3-540-74141-1_4. 39

[221] M T Rosenstein and A G Barto. *Supervised Actor-Critic Reinforcement Learning*. John Wiley & Sons, Inc., New York, NY, USA, 2004. 19, 58

[222] S Rosenthal, A K Dey, and M Veloso. How robots' questions affect the accuracy of the human responses. In *Proceedings of the IEEE Symposium on Robot and Human Interactive Communication*, pages 1137–1142, 2009. DOI: 10.1109/ROMAN.2009.5326291. 61

[223] Stephanie Rosenthal and Manuela Veloso. Modeling humans as observation providers using pomdps. In *RO-MAN, 2011 IEEE*, pages 53–58. IEEE, 2011. DOI: 10.1109/RO-MAN.2011.6005272. 61

[224] R. M. Ryan and E. L. Deci. Self-determination theory and the facilitation of intrinsic motivation, social development, and well-being. *American Psychologist*, 55(1):68–78, 2000. DOI: 10.1037/0003-066X.55.1.68. 6

[225] Paul E Rybski, Kevin Yoon, Jeremy Stolarz, and Manuela M Veloso. Interactive robot task training through dialog and demonstration. In *HRI '07: Proceedings of the ACM/IEEE international conference on Human-robot interaction*, pages 49–56, New York, NY, USA, 2007. ACM. DOI: 10.1145/1228716.1228724. 20, 40

[226] L M Saksida, S M Raymond, and D S Touretzky. Shaping robot behavior using principles from instrumental conditioning. *Robotics and Autonomous Systems*, 22(3/4):231, 1998. DOI: 10.1016/S0921-8890(97)00041-9. 55

[227] J Saunders, C Nehaniv, and K Dautenhahn. Teaching robots by moulding behavior and scaffolding the environment,. In *Proceedings of the ACM SIGCHI/SIGART Conference on Human-Robot Interaction*, pages 118–125, 2006. DOI: 10.1145/1121241.1121263. 10, 39

[228] Joe Saunders, Chrystopher L Nehaniv, Kerstin Dautenhahn, and Aris Alissandrakis. Self-imitation and environmental scaffolding for robot teaching. *International Journal of Advanced Robotics Systems*, 4(1):109–124, 2007. DOI: 10.5772/5703. 10

[229] J. Schmidhuber. A possibility for implementing curiosity and boredom in model-building neural controllers. In *International Conference on Simulation of Adaptive Behavior*, pages 222–227, 1991. 8

[230] Candace L Sidner, Cory D Kidd, Christopher Lee, and Neal Lesh. Where to look: a study of human-robot engagement. In *Knowledge Creation Diffusion Utilization*, IUI '04, pages 78–84. Mitsubishi Electric Research Labs and MIT Media Lab, ACM, 2004. DOI: 10.1145/964442.964458. 10

[231] O. Sigaud and J. Peters. *From Motor Learning to Interaction Learning in Robots*. Springer, 2010. DOI: 10.1007/978-3-642-05181-4. 81

[232] David Silver, J. Andrew Bagnell, and Anthony Stentz. Learning from demonstration for autonomous navigation in complex unstructured terrain. *The International Journal of Robotics Research*, 29(12):1565–1592, June 2010. DOI: 10.1177/0278364910369715. 34, 44, 79

[233] S Singh, A G Barto, and N Chentanez. Intrinsically motivated reinforcement learning. In *Proceedings of Advances in Neural Information Processing Systems 17*, 2005. 8

[234] B. F. Skinner. *Science and Human Behavior.* Colliler-Macmillian, 1953. 23

[235] Marjorie Skubic and Richard A Volz. Identifying contact formations from sensory patterns and its applicability to robot programming by demonstration. In *Intelligent Robots and Systems' 96, IROS 96, Proceedings of the 1996 IEEE/RSJ International Conference on,* volume 2, pages 458–464. IEEE, 1996. DOI: 10.1109/IROS.1996.570817. 26

[236] W D Smart and L P Kaelbling. Effective reinforcement learning for mobile robots. In *In Proceedings of the IEEE International Conference on Robotics and Automation,* pages 3404–3410, 2002. DOI: 10.1109/ROBOT.2002.1014237. 58

[237] William D Smart. Making reinforcement learning work on real robots. Ph.D. thesis, Department of Computer Science, Brown University, Providence, RI, 2002. 18, 19

[238] Dimitrios Stefanidis, Fikre Wang, James R Korndorffer Jr, J Bruce Dunne, and Daniel J Scott. Robotic assistance improves intracorporeal suturing performance and safety in the operating room while decreasing operator workload. *Surgical Endoscopy,* 24(2):377–382, 2010. DOI: 10.1007/s00464-009-0578-0. 70

[239] A. Steinfeld, T. Fong, D. Kaber, M. Lewis, J. Scholtz, A. Schultz, and M. Goodrich. Common metrics for human-robot interaction. In *Proceedings of the 1st ACM SIGCHI/SIGART Conference on Human-Robot Interaction,* HRI '06, pages 33–40, New York, NY, USA, 2006. ACM. DOI: 10.1145/1121241.1121249. 78

[240] Aaron Steinfeld, Terrence Fong, David Kaber, Michael Lewis, Jean Scholtz, Alan Schultz, and Michael Goodrich. Common metrics for human-robot interaction. In *HRI '06: Proceedings of the 1st ACM SIGCHI/SIGART conference on Human-robot interaction,* pages 33–40, New York, NY, USA, 2006. ACM. DOI: 10.1145/1121241.1121249. 69

[241] A Stern, A Frank, and B Resner. Virtual petz (video session): a hybrid approach to creating autonomous, lifelike dogz and catz. In *AGENTS '98: Proceedings of the Second International Conference on Autonomous Agents,* pages 334–335, New York, NY, USA, 1998. ACM Press. DOI: 10.1145/280765.280852. 23

[242] H. B. Suay, R. Toris, and S. Chernova. A practical comparison of three robot learning from demonstration algorithms. *International Journal of Social Robotics,* 2012. DOI: 10.1007/s12369-012-0158-7. 23, 58, 79

[243] Keith Sullivan. Multiagent hierarchical learning from demonstration. In *Proceedings of the Twenty-Second International Joint Conference on Artificial Intelligence-Volume Volume Three,* pages 2852–2853. AAAI Press, 2011. DOI: 10.5591/978-1-57735-516-8/IJCAI11-498. 38

[244] Keith Sullivan, Sean Luke, and Vittoria Amos Ziparo. Hierarchical learning from demonstration on humanoid robots. In *Proceedings of Humanoid Robots Learning from Human Interaction Workshop*, 2010. 38

[245] Richard S Sutton and Andrew G Barto. *Reinforcement Learning: An Introduction*. The MIT Press, Cambridge, MA, London, England, 1998. 2, 23, 55

[246] Richard S Sutton, Doina Precup, and Satinder Singh. Between mdps and semi-mdps: A framework for temporal abstraction in reinforcement learning. *Artificial Intelligence*, 112(1):181–211, 1999. DOI: 10.1016/S0004-3702(99)00052-1. 44

[247] Umar Syed, Michael Bowling, and Robert E Schapire. Apprenticeship learning using linear programming. In *Proceedings of the 25th International Conference on Machine Learning*, pages 1032–1039. ACM, 2008. DOI: 10.1145/1390156.1390286. 44

[248] Umar Syed and Robert E Schapire. A game-theoretic approach to apprenticeship learning. In *Advances in Neural Information Processing Systems*, pages 1449–1456, 2007. 44

[249] J. Takamatsu, K. Ogawara, H. Kimura, and K. Ikeuchi. Recognizing assembly tasks through human demonstration. *The International Journal of Robotics Research*, 26(7):641–659, July 2007. DOI: 10.1177/0278364907080736. 22

[250] A. Tenorio-Gonzalez, E. Morales, and L. Villase nor Pineda. Dynamic reward shaping: training a robot by voice. In *Advances in Artificial Intelligence–IBERAMIA*, pages 483–492, 2010. DOI: 10.1007/978-3-642-16952-6_49. 55

[251] Evangelos Theodorou, Jonas Buchli, and Stefan Schaal. Reinforcement learning of motor skills in high dimensions: A path integral approach. In *Robotics and Automation , 2010 IEEE International Conference on*, pages 2397–2403. IEEE, 2010. DOI: 10.1109/ROBOT.2010.5509336. 29

[252] A L Thomaz and C Breazeal. Reinforcement learning with human teachers: Evidence of feedback and guidance with implications for learning performance. In *Proceedings of the 21st National Conference on Artificial Intelligence*, 2006. 23, 58, 79

[253] A L Thomaz and C Breazeal. Teachable robots: Understanding human teaching behavior to build more effective robot learners. *Artificial Intelligence Journal*, 172:716–737, 2008. DOI: 10.1016/j.artint.2007.09.009. 23, 58, 79

[254] M. Tomasello. *The Cultural Origins of Human Cognition*. Harvard University Press, March 2001. 8

[255] R. Toris, D. Kent, and S. Chernova. The robot management system: A framework for conducting human-robot interaction studies through crowdsourcing. *International Journal of Human-Robot Interaction*, 2013. 79

[256] J.G. Trafton, M.D. Bugajska, B.R. Fransen, and R.M. Ratwani. Integrating vision and audition within a cognitive architecture to track conversations. In *Proceedings of the 3rd ACM/IEEE International Conference on Human Robot Interaction*, pages 201–208, 2008. DOI: 10.1145/1349822.1349849. 10

[257] E Tronik, H Als, L Adamson, and C Trevarthen. Communication and cooperation in early infancy: A description of primary intersubjectivity. In M Bullowa, editor, *Before Speech: The Beginning of Interpersonal Communication*, pages 389–450. Cambridge University Press, Cambridge, 1979. 7, 10

[258] K. M. Tsui, M. Desai, H. A. Yanco, H. Cramer, and N. Kemp. Using the "negative attitudes toward robots scale" with telepresence robots. In *Proceedings of the Performance Metrics for Intelligent Systems Workshop*, 2010. DOI: 10.1145/2377576.2377621. 70

[259] Ales Ude, Christopher G Atkeson, and Marcia Riley. Programming full-body movements for humanoid robots by observation. *Robotics and Autonomous Systems*, 47:93–108, 2004. DOI: 10.1016/S0921-8890(04)00040-5. 22

[260] Michael van Lent and John E Laird. Learning procedural knowledge through observation. In *K-CAP '01: Proceedings of the 1st International Conference on Knowledge Capture*, pages 179–186, New York, NY, USA, 2001. ACM Press. DOI: 10.1145/500737.500765. 40, 43

[261] D. Vasquez, T. Fraichard, and C. Laugier. Growing Hidden Markov Models: An incremental tool for learning and predicting human and vehicle motion. *The International Journal of Robotics Research*, 28(11-12):1486–1506, August 2009. DOI: 10.1177/0278364909342118. 31

[262] H Veeraraghavan and Manuela Veloso. Teaching sequential tasks with repetition through demonstration (short paper). In *Proceedings of the International Conference on Autonomous Agents and Multiagent Systems*, May 2008. 40

[263] Manuela Veloso, Felix Von Hundelshausen, and Paul E Rybski. Learning visual object definitions by observing human activities. In *Humanoid Robots, 2005 5th IEEE-RAS International Conference on*, pages 148–153. IEEE, 2005. DOI: 10.1109/ICHR.2005.1573560. 49

[264] Sethu Vijayakumar and Stefan Schaal. Locally weighted projection regression: An O(n) algorithm for incremental real time learning in high dimensional space. In *Proceedings of Seventeenth International Conference on Machine Learning*, pages 1079–1086, 2000. 28

[265] L. S. Vygotsky. *Mind in society: the development of higher psychological processes*. Harvard University Press, Cambridge, MA, 1978. 9

[266] A. R. Wagner. The role of trust and relationships in human-robot social interaction, 2009. 70

[267] Q. Wang, J. De Schutter, W. Witvrouw, and S. Graves. Derivation of compliant motion programs based on human demonstration. In *Robotics and Automation, 1996. Proceedings., 1996 IEEE International Conference on*, volume 3, pages 2616–2621 vol.3, 1996. DOI: 10.1109/ROBOT.1996.506557. 26

[268] J V Wertsch, N Minick, and F J Arns. Creation of context in joint problem solving. In B Rogoff and J Lave, editors, *Everyday Cognition: Its Development in Social Context*. Harvard University Press, Cambridge, MA, 1984. 9

[269] C. Wickens, J. Gordon, and Y. Liu. *An Introduction to Human Factors Engineering*. Pearson-Prentice Hall, New Jersey, 2004. 78

[270] Akiko Yamazaki, Keiichi Yamazaki, Yoshinori Kuno, Matthew Burdelski, Michie Kawashima, and Hideaki Kuzuoka. Precision timing in human-robot interaction: Coordination of head movement and utterance. *Ratio*, 08(1):131–139, 2008. DOI: 10.1145/1357054.1357077. 10

[271] P Zang, R Tian, A L Thomaz, and C Isbell. Batch versus interactive learning by demonstration. In *Proceedings of the International Conference on Development and Learning*, 2010. 54

[272] Brian D Ziebart, Andrew Maas, James (Drew) Bagnell, and Anind K Dey. Maximum entropy inverse reinforcement learning. In *Proceeding of AAAI 2008*, July 2008. 44

[273] Matthew Zucker and J Andrew Bagnell. Reinforcement planning: Rl for optimal planners. In *Robotics and Automation , 2012 IEEE International Conference on*, pages 1850–1855. IEEE, 2012. DOI: 10.1109/ICRA.2012.6225036. 45

[274] P Zukow-Goldring, M A Arbib, and E Oztop. Language and the mirror system: A perception/action based approach to cognitive development. *Cognition, Brain, Behavior*, pages 239–272, 2005. 7, 9

Authors' Biographies

SONIA CHERNOVA

Sonia Chernova is an Assistant Professor of Computer Science and Robotics Engineering at Worcester Polytechnic Institute and the director of the Robot Autonomy and Interactive Learning (RAIL) lab. She earned B.S. and Ph.D. degrees in Computer Science from Carnegie Mellon University in 2003 and 2009, and was a Postdoctoral Associate at the MIT Media Lab prior to joining WPI. Dr. Chernova's research is focused on interactive machine learning, adjustable autonomy, crowdsourcing, and human-robot interaction. She has received funding support from NSF, ONR, and DARPA, including an NSF CAREER award on Learning from Demonstration in 2012.

ANDREA L. THOMAZ

Andrea L. Thomaz is an Associate Professor of Interactive Computing at the Georgia Institute of Technology. She directs the Socially Intelligent Machines lab, which is affiliated with the Robotics and Intelligent Machines (RIM) Center and with the Graphics Visualization and Usability (GVU) Center. She earned a B.S. in Electrical and Computer Engineering from the University of Texas at Austin in 1999, and Sc.M. and Ph.D. degrees from MIT in 2002 and 2006. Dr. Thomaz has published in the areas of Artificial Intelligence, Robotics, and Human-Robot Interaction. She has received recognition as a young leader in her field, receiving an ONR Young Investigator Award in 2008, and an NSF CAREER award in 2010. Her work has been featured on the front page of *The New York Times*, on *NOVA Science Now*, she was named one of *MIT Technology Review*'s Top 35 under 35 in 2009, and on *Popular Science Magazine*'s Brilliant 10 list in 2012.